Radionuclides and Heavy Metals in the Environment

Series Editors:

Dharmendra K. Gupta
Ministry of Environment, Forest and Climate Change
Indira Paryavaran Bhavan, Jorbagh Road
Aliganj, New Delhi, India

Clemens Walther
Gottfried Wilhelm Leibniz Universität Hannover
Institut für Radioökologie und Strahlenschutz (IRS)
Hannover, Germany

More information about this series at http://www.springer.com/series/16207

Dharmendra K. Gupta • Soumya Chatterjee
Clemens Walther

Editors

Lead in Plants and the Environment

 Springer

Editors
Dharmendra K. Gupta
Ministry of Environment, Forest and
Climate Change
Indira Paryavaran Bhavan, Jorbagh Road
Aliganj, New Delhi, India

Soumya Chatterjee
Defence Research Laboratory
Defence Research & Development
Organisation
Tezpur, Assam, India

Clemens Walther
Gottfried Wilhelm Leibniz Universität
Hannover
Institut für Radioökologie und
Strahlenschutz (IRS)
Hannover, Germany

ISSN 2524-7409   ISSN 2524-7417   (electronic)
Radionuclides and Heavy Metals in the Environment
ISBN 978-3-030-21640-5   ISBN 978-3-030-21638-2   (eBook)
https://doi.org/10.1007/978-3-030-21638-2

This Springer imprint is published by the registered company Springer Nature Switzerland AG
The registered company address is: Gewerbestrasse 11, 6330 Cham, Switzerland

Preface

Lead (Pb) is a metal utilized by humans for many thousands of years. Metallic Pb globules manufactured in 6400 BC were discovered at Çatalhöyük (presently in Republic of Turkey). Pb is a bluish-white lustrous metal, which is very soft, highly malleable, and ductile, and is a relatively poor electric conductor. Metal is resistant to corrosion but tarnishes upon exposure to air. In nature, it is typically found as minerals, in combination with other elements.

A total of 49 isotopes of Pb were recorded till date with four stable isotopes (^{204,206,207,208} Pb). Among the stable isotopes, only ^{204}Pb is a primordial nuclide, and not a radiogenic one. The three other stable isotopes, ^{206,207,208}Pb, are the endpoints of three decay chains, i.e., uranium, actinium, and thorium series, respectively. ^{205}Pb and ^{202}Pb are the longest-lived radioisotopes with a half-life of approximately 15.3 million years and 53,000 years, respectively. The radiologically most relevant radioactive nuclide ^{210}Pb is part of the ^{238}U series and has a half-life of 22.3 years. ^{210}Pb is suitable for studying the chronology of sedimentation on time scales shorter than 100 years. Anthropogenic activities, like combustion of coal, are one of the major sources of ^{210}Pb in the atmosphere, but ^{210}Pb also occurs naturally since it is a progeny of the radioactive noble gas radon (^{222}Rn) emanating from soil air due to the omnipresent uranium.

Pb is a microelement with no known physiological function but found in trace amounts in all biotic resources, e.g., in soil, water, plants, and animals. Pb is a toxic element, pollution of which may come from various sources. In the environment, nearly 98% of stable Pb originates from paints, petrochemicals, pipes and supply systems, etc. Routes of Pb poisoning may be through consumption of contaminated food and water, breathing contaminated air from cigars and automobile exhausts, and using uncleaned adulterated hands/face where individual health and hygiene issues are compromised. However, usually, Pb is not absorbed through skin.

Recent extensive work on ^{210}Pb radioisotope for examining plant uptake, where, mostly, artificial spiking of the metal in the soil and observing its consequent absorption in plant and soils. Usually, Pb forms complexes with soil particles, and a very small amount or fractions are easily available for plants. Despite its lack of essential function in plants, Pb is taken up mostly through the roots from soil solutions at rhizosphere level, which may cause the entry of Pb into the food chain. It is also reported that uptake of Pb by roots occurs mainly through apoplastic pathway or via Ca^{2+} absorbent channels. Pb in soil and its uptake by plants depends on several factors, like soil pH, soil particle size, soil moisture, cation-exchange capacity, presence of other (in)organic substances (including humus), root structure and rhizosphere, root exudates, and root mycorrhizal properties. Once Pb enters into the plants through root, initially, it is getting deposited at root cells. However, reports also suggest that negatively charged root cell walls adsorb Pb. Accumulation of Pb in plants renders phytotoxic symptoms, disturbing morphological, physiological, and biochemical functions, like inhibition of ATP production, lipid peroxidation, and DNA damage by overproduction of reactive oxygen species (ROS). However, monitoring Pb remobilization and related secondary pollution and effective, environment-friendly remediation measures to reduce Pb pollution is the need of the hour.

The main features of this volume are interrelated to how Pb enters into the environment and its translocation from soil to plants and into the food chain. Chapters 1 and 2 deal with the analytical methods for determining Pb both in environmental and in biological samples and also the effect of radioisotopic lead behavior in plants and environment and distribution of radioactive Pb and its distribution in environment through modelling application. Chapters 3 and 4 focus on Pb exposure to humans via agroecosystem and its consequences. Chapters 5 and 6 focus on how Pb behaves in soil plant system and how it uptakes in plants. Chapters 7 and 8 emphasize on how Pb reacts on physiological and biochemical changes in plants with reference to different plant enzymes and photosynthetic apparatus. Last but not least, Chaps. 9 and 10 present the biological strategies of lichens symbionts, under Pb toxicity, and how Pb pollution is going to remediate via phytoremediation. The material composed in this volume will bring in-depth holistic information on Pb (both stable and radioactive) uptake and translocation and its toxicity in plants and effect on human health and phytoremediation strategies.

Drs. Dharmendra K. Gupta, Soumya Chatterjee, and Prof. Clemens Walther individually acknowledge all authors for contributing their valuable time, information, and interest to bring this book into its current form.

New Delhi, India   Dharmendra K. Gupta
Tezpur, Assam, India   Soumya Chatterjee
Hannover, Germany   Clemens Walther

Contents

About the Editors

Dharmendra K. Gupta is Director at Ministry of Environment, Forest and Climate Change, Indira Paryavaran Bhavan, Jorbagh Road, Aliganj, New Delhi, India. He already published more than 90 refereed research papers/review articles in peer-reviewed journals and in books and also edited 16 books. His field of research includes abiotic stress caused by radionuclides/heavy metals and xenobiotics in plants, antioxidative system in plants, and environmental pollution (radionuclides/heavy metals) remediation through plants (phytoremediation).

Soumya Chatterjee is Senior Scientist and Head of the Department of Biodegradation Technology at Defence Research Laboratory (DRDO) at Tezpur, Assam, India. His area of research includes microbial biodegradation, abiotic stress in plants, bioremediation and phytoremediation, wastewater bacteriophages, sanitation, and metagenomics. He has already published more than 60 refereed research papers/review articles and book chapters in peer-reviewed journals/books (including edited books and journal special issues).

Clemens Walther is Professor of Radioecology and Radiation Protection and Director of the Institute for Radioecology and Radiation Protection at the Leibniz Universität Hannover, Germany. He published more than 100 papers in peer-reviewed journals. His field of research is actinide chemistry with a focus on solution species and formation of colloids and ultra-trace detection and speciation of radionuclides in the environment by mass spectrometry and laser spectroscopy.

Major Analytical Methods for Determining Lead in Environmental and Biological Samples

Jozef Sabol

Abstract Lead (Pb) is an element which is found in nature where out of its known 49 isotopes the most abundant are four: ^{208}Pb (52%), ^{206}Pb (24%), ^{207}Pb (22%) and ^{204}Pb (less than 2%). The increased concentration of lead in the environment is mainly due to some human activities. This includes use of petrol in transport vehicles and releases from industrial and other installations and facilities. From the contaminated environment where soil, water, air, animals and plants always contain certain concentration of lead, the nuclide can find a way into the human organism by inhalation and ingestion. This results in some health effects which, in the case of higher intake, may be extremely poisoning and dangerous. Chronic lead intoxication has been linked to Alzheimer's disease. Lead, like many heavy elements, tends to accumulate in bone. Therefore, analysis of the lead presence in environmental and biological samples is an important prevention measure against harmful consequences which must be minimised and in accordance with the set standards and limits. The chapter deals with some specific methods recommended for determining lead in various samples. Special attention is paid to the description of XRF and PIXE methods and especially methods based on atomic spectroscopy, namely absorption and emission spectroscopy as well as atomic fluorescence methods. While the first two methods are essentially considered as non-destructive, the atomic spectroscopy method falls into the category of destructive methods.

Keywords Lead (Pb) · Health effects · Determination · Atom absorption spectroscopy · Atomic emission spectroscopy · Atomic fluorescence spectrometry · Environmental samples · Biological samples

J. Sabol (✉)
Department of Crisis Management, Faculty of Security Management, PACR,
Prague, Czech Republic
e-mail: sabol@polac.cz

© Springer Nature Switzerland AG 2020
D. K. Gupta et al. (eds.), *Lead in Plants and the Environment*, Radionuclides
and Heavy Metals in the Environment,
https://doi.org/10.1007/978-3-030-21638-2_1

1

1 Introduction

Lead (Pb) is a blue-gray malleable metal found in Group 14 (IV A) periodic table of elements. Natural lead is a mixture of predominantly four stable isotopes: ^{208}Pb, ^{206}Pb, ^{207}Pb and ^{204}Pb. Altogether, there are 49 known lead isotopes, the occurrences of the four of them mentioned above are most abundant: ^{208}Pb (52%), ^{206}Pb (24%), ^{207}Pb (22%) and ^{204}Pb (less than 2%). Of these four isotopes of lead, the only ^{204}Pb is non-radiogenic (it is not the product of radioactive decay, it originates outside the decaying series) and its presence on Earth is on Earth is unvarying. The other three are radiogenic final isotopes of the disintegration series: ^{206}Pb is the final disintegration product of ^{238}U (uranium ^{238}U decay chain), ^{207}Pb is the final product of ^{235}U (uranium ^{235}U decay chain), and ^{208}Pb is the ending product of ^{232}Th (Table 1). It is possible to distinguish with different isotopic composition whether it is a natural or anthropogenic source. This is possible if all pollution sources are characterised by their ratio of lead isotopes and pollutants therefore have their own specific isotopic composition. For example, lead released during combustion processes shows a different isotopic composition of ^{206}Pb/^{207}Pb from other sources of pollution. Also, the isotopic composition of lead emitted into the atmosphere in metallurgical processes corresponds to the isotopic composition of the original materials.

Lead was known and used since prehistoric times. Exposure to lead has been consequently increased mainly because the environment is more and more contaminated by this element. Although acute lead poisoning has become sporadic, chronic exposure to low levels of lead is still considered to be a public health issue (Shilu et al. 2000). Lead intake can lead to a variety of adverse health impacts All over the word; the relevant standards for lead emissions have become increasingly stringent because of new findings about its possible health impacts. The associated regulations require monitoring the situation in order to keep the level and lead concentration below the limits and action levels set by national regulatory authorities.

Lead had previously been mainly introduced into the waters by road and vegetation flushes from the immediate vicinity of busy roads. The exhaust gas contained lead as the decomposition products of tetra-alkyl, which was a common anti-knock additive for gasoline. To a lesser degree lead and lead alloy plants also contribute to contamination. It also gets into the water from the lead pipe. The tetra-alkyl compounds are very volatile and easily pass into the atmosphere when aerating the water. It accumulates in bones and other tissues with age. It interferes with red blood cell enzymes and may cause death at higher doses, causing mental retardation. Often enough lead (200 mg kg^{-1}) in the grass along the highways could even kill cattle.

The nuclides produced as results of the decay of natural radionuclides with very long half-life such as ^{238}U (half-life 4.5 10^9 years), ^{235}U (half-life 0.7 10^9 years) and ^{232}Th (half-life 14 10^9 years) are radioactive and continue to form more radioactive nuclides until a non-radioactive nuclide is formed. A uranium-radium decay chain begins with ^{238}U and ends with the stable ^{206}Pb after going through 18 intermediate steps. Uranium-235 is at the beginning of the uranium-actinium decay chain leading via 15 radionuclides to ^{207}Pb. With ten intermediate states, the thorium decay chain

Table 1 Three main decay chains (or families) are observed in nature, commonly called the thorium series, the radium or uranium series, and the actinium series, representing three of these four classes, and ending in three different, stable isotopes of lead (*A*, mass number corresponding to number of protons and neutrons; *Z* atomic or proton number; *N* number of neutrons)

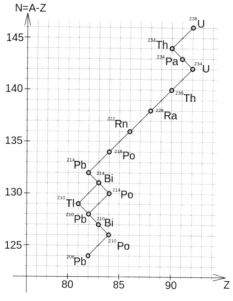

Nuclide	Half-life	Decay	
^{238}U	4.468×10^9 y	α	
^{234}Th	24.10 d	β⁻	
^{234m}Pa	1.17 min	β⁻	
^{234}U	2.455×10^5 y	α	
^{230}Th	7.538×10^4 y	α	
^{226}Ra	1600 y	α	
^{222}Rn	3.8235 d	α	
^{218}Po	3.10 min	α	
^{214}Pb	26.8 min	β⁻	
^{214}Bi	19.9 min	β⁻	α
^{214}Po	164.3×10^{-6} s	(0.02 %) α	
^{210}Tl	1.30 min	β⁻	
^{210}Pb	22.20 y	β⁻	
^{210}Bi	5.012 d	β⁻	
^{210}Po	138.376 d	α	
^{206}Pb	Stable		

Uranium series

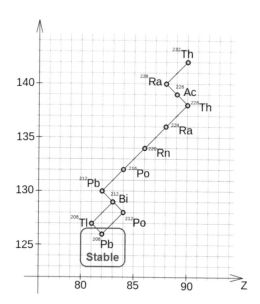

Nuclide	Half-life	Decay	
^{232}Th	1.405×10^{10} y	α	
^{228}Ra	5.75 y	β⁻	
^{228}Ac	6.15 h	β⁻	
^{228}Th	1.9116 y	α	
^{224}Ra	3.66 d	α	
^{220}Rn	55.6 s	α	
^{216}Po	0.145 s	α	
^{212}Pb	10.64 h	β⁻	
^{212}Bi	60.55 min	β⁻	α
^{212}Po	0.299×10^{-6} s	(35.94 %) α	
^{208}Tl	3.053 min	β⁻	
^{208}Pb	Stable		

Thorium series

(continued)

Table 1 (continued)

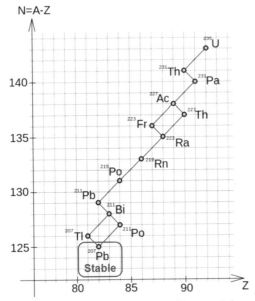

Nuclide	Half-life	Decay	
^{235}U	7.04×10^8 y	α	
^{231}Th	25.52 h	β⁻	
^{231}Pa	3.276×10^4 y	α	
^{227}Ac	21.772 y	β⁻	α
^{227}Th	18.68 d		(1.38 %)
^{223}Fr	22.00 min	α	β⁻
^{223}Ra	11.43 d	α	
^{219}Rn	3.96 s	α	
^{215}Po	1.781×10^{-3} s	α	
^{211}Pb	36.1 min	β⁻	
^{211}Bi	2.14 min	β⁻	α
^{211}Po	0.516 s		(99.72 %)
^{207}Tl	4.77 min	α	β⁻
^{207}Pb	Stable		

Uran-actinium series

starting with ^{232}Th and ending at ^{208}Pb is the shortest. The daughter nuclides arising from the disintegration of naturally occurring ^{238}U, ^{235}U and ^{232}Th are all radioactive and therefore disintegrate until the last one which is stable lead, namely ^{206}Pb, ^{207}Pb and ^{208}Pb, respectively.

2   Health Effects

While the use of lead has been greatly reduced in developed countries, it is still used widely in developing countries because it has some unique useful properties suitable and required in a number of various applications. The continued use of lead has caused its levels to raise worldwide, posing serious threats not only to the environment but also to the humans living in it. Lead can cause some damage in every organ and tissue in human body, Exposure to high lead levels can severely affect the brain and kidneys and eventually cause death. High levels of exposure can lead to miscarriage in pregnant women. Lead is also widely thought to be cancerogenic. Toxic effects of lead have been observed especially on the renal, reproductive and nervous system. Therefore, some the techniques were developed for treating lead toxicity. Some more information about the recent progress in this area is given in the review

Table 2 Toxicology of lead (adopted from Das and Grewal 2011)

Form entering body	Major route of absorption	Distribution	Major clinical effects	Key aspects of mechanism	Metabolism and elimination
Inorganic lead oxides and salts	Gastrointestinal, respiratory	Soft tissues; redistributed to skeleton (>90% of adult body burden	CNS deficits; peripheral neuropathy; anaemia; nephropathy; hypertension; reproductive toxicity	Inhibits enzymes; interferes with essential cations; alters membrane structure	Renal (major); faeces and breast milk (minor)
Organic (tetraethyl lead	Skin, gastrointestinal, respiratory	Soft tissues, especially liver, CNS	Encephalopathy	Hepatic de alkylation (fast) tri alky metabolites (slow) dissociation to Pb	Urine and faeces (major); sweat (minor)

(Wani et al. 2015). An overview of possible health effects caused by lead which entered human body through various routes is given in Table 2.

In general, it has been widely accepted that lead is a probable human carcinogen. Lead can affect every organ and system in the body. Exposure to high lead levels can severely damage the brain and kidneys and ultimately cause death. In pregnant women, high levels of exposure to lead may cause miscarriage. Internationally for lead regulatory limits have been introduced corresponding to15 parts per billion (ppb) in drinking water and 0.15 µg per cubic meter in air.

Since lead is an element that occurs naturally in the earth, trace amounts of lead may occur in the foods we eat and the water we drink. Regulatory authorities in most countries are trying to control population exposure to lead from various sources. The main sources of lead entering human body include the following:

- Inhalation of lead dust which originates from lead-based paint and lead-contaminated soil.
- Touching by hands other objects contaminated with lead dust and then putting them into mouths.
- Consuming food, candy or water contaminated by lead.
- By means of dishes or glasses that contain lead.
- From colour additives in paints and cosmetics (special case may be children playing with toys that contain lead paint).

In addition to occupationally related exposure by some workers, the most sensitive group among population represent children, which may be affected by lead due to many different situations (Fig. 1).

Fig. 1 The main routes of
lead entering children's
bodies

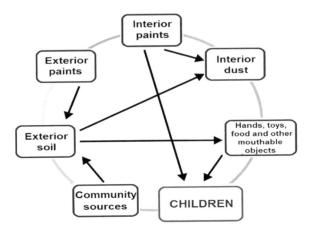

3   Overview of Basic Analytical Methods for Determining Lead

There are several methods used for the monitoring and identification of the occurrence of lead in analysed samples. The techniques differ as to sample preparation and treatment as well as to instrumentation used.

3.1   X-Ray Fluorescent Spectroscopy and PIXE Method

The principles of both methods are similar. In case of X-ray fluorescence (XRF) method, an incident X-ray photon removes an orbital electron leaving a hole which is filled by an electron from an outer shell which results in the emission of a photoelectron and a characteristic photon with an energy inherent to the type of the atom of a nuclide under examination. The PIXE (particle induced X-ray emission) method is essentially the same, only instead incident X-ray photons charge particles (in most cases protons) are used. The process is illustrated in Fig. 2.

X-ray photons or charged particles excite inner electron shells of the sample material. The ejected electron leaves the atom of the target element as a photoelectron. Other electrons fill the gap and give off large amounts of energy in the form of characteristic X-rays, which are detected, their energy can thus identify the element, and the intensity of X-rays identifies concentration of the element in the sample. In general, incident photons may be produced either by an X-ray tube or a radionuclide emitting soft gamma radiation.

As a source of primary radiation, the XRF analysers can use either an X-ray tube (Fig. 3a) or a suitable radioactive excitation source (Fig. 3b) such as ^{57}Co and ^{109}Cd.

In addition to laboratory XRF analysers, usually equipped with radioactive sources, there are several various types of portable analysers (Fig. 4) designed to

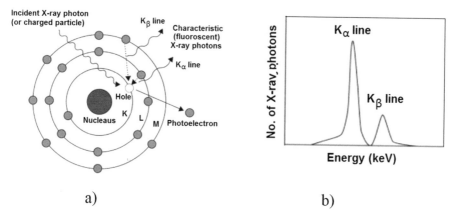

Fig. 2 The principle of the mechanism of XRF (and PIXE) method; (**a**) the generation of characteristic photons; (**b**) their energy distribution (Based on Motohiro et al. 2015)

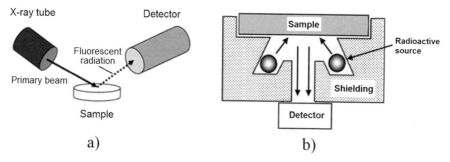

Fig. 3 Basic arrangement of an XRF analyser using (**a**) an X-ray tube and (**b**) a radioactive source

monitor the detection of heavy metals (including Pb) in soil, paint, toys, and so on. These handheld XRF analysers can also be used for specific material identification and hazardous material analysis, and for metal alloy identification, consumer goods screening, compliance screening and many other analysis needs. These analysers serve also for checking compliance with the standards and limits introduced by the relevant regulatory authorities,

An X-ray spectrographic technique known as the *PIXE method*, can be used to analyse solid, liquid or aerosol filter samples in a non-destructive, simultaneous elemental way. The X-ray photons are initiated when energetic protons excite target atoms in the inner shell of electrons. When these inner shell electrons are subsequently expelled, they produce X-rays whose energies are emitted when the resulting vacancies are again filled. These vacancies are unique to the sample elements being analysed, with the number of X-rays emitted being proportional to the mass of the corresponding sample element.

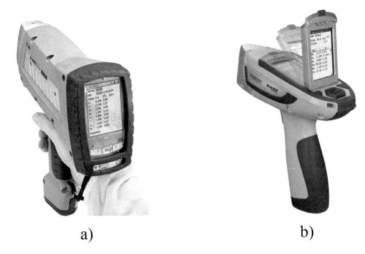

a)   b)

Fig. 4 Some examples of portable XRF analysers, (**a**) Oxford X-MET5000 Handheld XRF Analyzer (Oxford 2019) and (**b**) Thermo scientific XRF analyser (Thermo 2019)

The PIXE method for the most part uses protons to generate X-rays in a sample, the probability of which depends on the proton energy (in MeV) as well as on the total number of incident protons. This number can be expressed as proton current (in mA): the greater the proton current, the greater the probability for X-ray production. As the proton energy changes, the probability for X-ray production also changes. Both of these factors must be accurately known in order to perform a correct quantitative analysis. When protons interact to produce X-rays, each collision in turn transfers kinetic energy from the mobile proton to the immobile target atom. While each collision produces a small amount of energy, as the collisions and resulting energy increase, they eventually reduce the proton's energy as well as its ability to generate X-rays, and in the end the proton becomes immobile. The instrument calibration is carried out at specific proton energy, Data on proton energy loss is necessary to calibrate instruments for performing accurate quantitative analysis. Since instrument calibration is carried out at specific proton energy, information of the proton energy loss is indispensable for quantitative analysis. The PIXE technique has been used in the monitoring of lead and other metals. The range of Pb concentrations in human rib bone was found to be in the range of 1.4–11.5 μg g^{-1} for the trabecular surface by PIXE (Deibel et al. 1995).

3.2 Atomic Spectroscopy

In general, atomic spectroscopy (spectrometry) represents the determination of elemental composition based on the evaluation of electromagnetic radiation absorbed and emitted. The analysis of the electromagnetic spectrum of elements, called

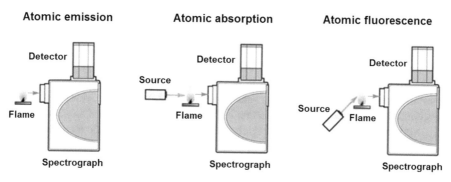

Fig. 5 The illustration of all three techniques of atomic spectroscopy based on atomic emission, atomic absorption and atomic fluorescence

optical atomic spectroscopy, includes three techniques: atomic absorption, atomic emission and atomic fluorescence versions (Fig. 5). The excitation and decay to the ground state is involved in all three fields of atomic spectroscopy. Either the energy absorbed in the excitation process, or the energy emitted in the decay process is measured and used for analytical purposes.

Atomic absorption measures the amount of light of a certain (resonant) wavelength which is absorbed as it penetrates through a cloud of atoms; the greater the number of atoms the greater the amount of light absorbed. This measurement can give a quantitative determination of the amount of analysed element in the sample. The specific light source emitting photons of suitable wavelength can determine quantitatively individual elements among other elements. This technique is fast and accurate, making atomic absorption a popular method for determining the amount of lead as well as other metals in any given substance.

Atomic emission spectrometry is a technique in which a high energy, thermal environment such as an electrical arc, a flame, or even plasma is applied to a sample in order to produce excited atoms which can emit light. The resulting emission spectrum consists of the discrete wavelengths (emission lines) which can also be used for qualitatively identifying an element.

The third field of atomic spectroscopy is *atomic fluorescence*. This method incorporates aspects of both atomic absorption and atomic emission modality. Like atomic absorption, ground state atoms created in a flame are excited by focusing a beam of light into the atomic vapor. Instead of looking at the amount of light absorbed in the process, however, the emission resulting from the decay of the atoms excited by the source light is measured.

Atomic fluorescence spectrometry takes characteristics from atomic absorption as well as atomic emission. Like atomic absorption, a beam of light is shone into atomic vapour, exciting ground state atoms created in a flame. Unlike it, however, the emission from the decay of the atoms excited is measured.

3.2.1 Atomic Absorption Spectroscopy

Atomic absorption spectroscopy (spectrometry), AAS, is a spectrometric analytical method used to determine both trace and significant concentrations of individual elements in an environmental or biological sample. The method can analyse over 60 elements of the periodic table with sensitivity from about 0.01–100 ppm. The greatest boom was recorded in the 1960s and 1980s, when it was one of the most sensitive and widely used instrumental analytical techniques. In forensic chemistry, it is mainly used to detect heavy metals and ammunition residues.

The analytical sample solution is fogged, and the resulting aerosol is introduced into a flame or graphite atomiser, where the solution is immediately evaporated and the chemical bonds in the molecules of the present compounds are broken. At the same time, the atomisation conditions are chosen so that the largest possible population of measured atoms remains in the neutral state and does not ionise to form charged particles.

A beam of light passes through a flame from a special discharge lamp whose photons are absorbed when they meet the atoms of the analysed element and the atom of the element passes into the relevant excited state. This leads to a decrease in the intensity of the transmitted light, and the loss is given by the Lambert–Beer law in the following form

$$I = I_0 e^{-(k.n.l)}$$

where I_0 is the intensity of the exciting radiation, I is the intensity of the radiation after passing through the absorbing environment (flame), k is the atomic absorption coefficient for the specified absorption line, n is the number of atoms of the analysed element in the volume unit and l is the length of the absorption layer.

In practice, the logarithm of the attenuation of light energy called absorbance (A) is used as the measure corresponding to the concentration of the nuclide monitored. The absorbance is expressed by the relation

$$A = \log \frac{I_0}{I} = 2.303 \left(k.n.l \right)$$

A very simple linear dependence on the atomic concentration of the measured element is then valid for absorbance. Therefore, all AAS spectrometers indicate the measured results in terms of absorbance units reflecting the actual measured light transmittance. Individual parts of a common atomic absorption spectrometer are shown in Fig. 6.

The atomic emission spectrometry is suitable for the assessment of lead concentration in various waters. Results of some measurements in Iran are presented in Table 3 (Dadfarnia et al. 2001). Another table (Table 4) illustrates the use of atomic absorption spectrometry in determining lead concentration in tea leaves, mixed Polish herbs and oriental tobacco leaves (Chwastowska et al. 2008). The same authors (Chwastowska et al. 2008) measured also lead concentration in soil, street dust and grass with the results of 67.6, 21.3 and 2.0 $\mu g\ g^{-1}$, respectively.

Fig. 6 Block diagram of a typical atomic absorption spectrometer (Based on Alshana 2007)

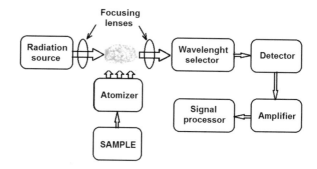

Table 3 Determination of lead in natural waters; sample volume: 1000 mL (pH = 3); eluent: 5 mL of HNO3 (4 M)

Samples	Concentration ($\mu g\ L^{-1}$)
Tap water	2.4 ± 0.1
River water	2.8 ± 0.1
Well water	2.0 ± 0.1
Spring water	3.2 ± 0.1

Table 4 Results of analysis of the concentration of lead together with Cu and Cd determined in three different samples

Sample	Pb ($\mu g\ g^{-1}$)	Cu ($\mu g\ g^{-1}$)	Cd ($\mu g\ g^{-1}$)
Tea leaves	1.78 ± 0.24	20.4 ± 1.5	0.030 ± 0.004
Mixed Polish herbs	2.16 ± 0.23	7.77 ± 0.53	0.199 ± 0.015
Oriental tobacco leaves	4.91 ± 0.80	14.1 ± 0.50	1.12 ± 0.12

3.2.2 Atomic Emission Spectroscopy

Atomic emission spectroscopy relies on the principle that when light or heat is applied to a molecule, it gets excited and moves to a higher energy level, making it unstable. The excited molecule then jumps to a lower energy level, thereby emitting photons of characteristic energy. The emitted wavelengths are then recorded in the emission spectrometer.

Similarly, to atomic absorption spectroscopy, the sample should be transformed into free atoms. This is normally achieved using a high-temperature excitation source. Liquid samples are dispersed and led in the excitation source by a flowing gas. Solid samples can be introduced into the source by slurry or by laser ablation of the solid sample in a gas stream. Solids can also be directly vaporised and excited by a spark between electrodes or by a laser pulse. The excitation source must desolvate, atomise, and excite the analyte atoms. The excitation sources include flame, inductively coupled plasma, laser-induced plasma, direct-current plasma, microwave-induced plasma and spark or arc. The basic scheme of a standard atomic emission spectrometer is in Fig. 7.

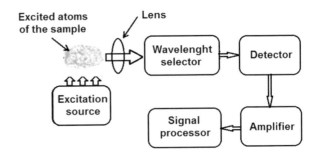

Fig. 7 Principal arrangement of an atomic emission spectrometer

3.2.3 Atomic Fluorescent Spectroscopy

An X-ray region in the electromagnetic spectrum is that which lies between the wavelengths of 0.01 and 10 nm. X-rays are produced, for example, in elementary reactions particles, in the decay of radioactive elements and other processes. Secondary X-rays are also produced by exposing the sample with appropriate (primary) X-rays. This phenomenon is called X-ray fluorescence and is the basis for qualitative and quantitative X-ray fluorescence spectroscopy (XRF—X-ray fluorescence) analytical techniques. Upon absorption of the primary photon by the electron in the inner shell of the atom, the release of this electron (photo effect) and the formation of an electron hole occur. This hole is filled with an electron jumping from higher energy level. In doing so, the secondary (fluorescent) X-rays are released, the spectrum of which is of a line character and characteristic of the element being analysed.

As a source of the fluorescent radiation, usually an X-ray tube or suitable radioactive sources are used. The X-ray source consists of an evacuated tube, tungsten fibre cathode, water-cooled target anode (Ca, Rh, Pd, Ag, W), supplied voltage source (5–80 kV) and a beryllium window. A radioactive source is an appropriate radionuclide emitting gamma radiation which is usually of lower intensity compared to X-ray tube. These sources are especially useful in smaller and portable spectrometers. Characteristic parameters of some suitable radionuclides used in XRF spectrometers are presented in Table 5.

4   Conclusion

Lead is clearly among the toxic elements. Historically, overuse of lead is one of the factors that have contributed to the extinction of the Roman Empire. The main contribution to the excessive intake of lead was due to the use of lead acetate as a sweetener. At present, lead is a ubiquitous environmental contaminant due to the use of lead in drinking water installations, in the production of paints, as an additive in gasoline and its other industrial applications. Lead penetrates into the body mainly in the bone, and a certain amount is found in the blood. Even traces of lead in the environment and food can lead to subsequent severe illnesses, as lead is accumulated in the body. Therefore, the instrumentation for the measurement and monitoring of

Table 5 Radionuclides suitable for the use in XRF spectrometry

Radionuclide	Half-life	Photon energy (keV)	Elements excited
Fe-55	2.7 years	5.9	Na up to V
Cd-109	453 days	22, 88	Cu up to Mo (K) Sn up to U (K)
Am-241	433 years	59.5	Sn up to Tm (K)
Co-57	272 days	122	Ta up to U (K)

lead concentration in specific environmental and biological samples is so important. For the time being, there are some gaps in controlling lead and in adoption of consistent regulations and standards where the situation must be improved.

References

Alshana U (2007) Atomic absorption and emission spectrometry. Online: https://slideplayer.com/slide/8889467/. Accessed 5 Jan 2019

Chwastowska J, Skwara W, Sterliñska E, Dudek J, Dabrowska M, Pszonicki L (2008) GF AAS determination of cadmium, lead and copper in environmental materials and food products after separation on dithizone sorbent. Chem Anal 53:887–894

Dadfarnia S, Haji Shabani AM, Dehgan Shirie H (2001) Determination of lead in different samples by atomic absorption spectrometry after preconcentration with dithizone immobilized on surfactant-coated alumina. Bull Korean Chem Soc 23:545–548

Das SK, Grewal AS (2011) A brief review: heavy metal and their analysis. Int J Pharm Sci Rev Res 11:13–18

Deibel MA, Savage JM, Robertson JD, Ehmann WD, Markesbery WR (1995) Lead determinations in human bone by particle induced X-ray emission (PIXE) and graphite furnace atomic absorption spectrometry (GFAAS). J Radioanal Nucl Chem 195:83–89

Motohiro U, Wada T, Sugiyama T (2015) Applications of X-ray fluorescence analysis (XRF) to dental and medical specimens. Jap Dent Sci Rev 51:2–9

Oxford (2019) Oxford X-MET5000 Handheld XRF Analyzer. Online: https://www.metalsanalyzer.com/oxford-instruments.html. Accessed 30 Jan 2019

Shilu T, von Schirnding YE, Prapamontol T (2000) Environmental lead exposure: a public health problem of global dimensions. Bull World Health Organ 78:1068–1077

Thermo (2019) Thermo Scientific Niton XL3t GOLDD XRF Analyzer. Online: https://www.azom.com/equipment-details.aspx?EquipID=443. Accessed 30 Jan 2019

Wani AL, Ara A, Usmani JA (2015) Lead toxicity: a review. Interdiscip Toxicol 8:55–64

Environmental Distribution and Modelling of Radioactive Lead (210): A Monte Carlo Simulation Application

Fatih Külahcı

Abstract The abundance of lead element with an atomic number 82 is 1.03×10^{-8} % in the Solar System, 14 mg kg^{-1} in the Earth's surface and 3×10^{-5} mg L^{-1} in the oceans. The most dangerous radioisotope of lead is the ^{210}Pb, which has a half-life of 22.26 years and gamma energy of 46.5 keV. Modelling is one of the most effective ways of appreciation about the distribution effects and transport of the elements to the earth. It has a wide range of content from pure differential equations to spatial analysis calculations. In this section, the modelling with Monte Carlo Simulation method of the environmental distribution of the lead can be found. The Monte Carlo Simulation method on ^{210}Pb data lead to concentrations for future times. In addition, models of auto regressive integrated average (ARIMA), generalized autoregressive conditional heteroscedastic (GARCH) and autoregressive conditional heteroscedastic (ARCH) are obtained to determine the environmental distribution characteristics of the lead. The proposed simulation methodologies can also be used successfully for other variables other than lead.

Keywords Lead 210 · Monte Carlo simulation · ARIMA · Modelling · Probability distribution · GARCH · Forecasting

1 Introduction

Lead (Pb) is one of the most important pollutants and its mining, production and recovery are possible in the environment its use in technology, and especially, in shielding, battery and toy industry. Pb is available in a high proportion within the

F. Külahcı (✉)
Science Faculty, Physics Department, Nuclear Physics Division, Firat University,
Elazig, Turkey
e-mail: fatihkulahci@firat.edu.tr

© Springer Nature Switzerland AG 2020
D. K. Gupta et al. (eds.), *Lead in Plants and the Environment*, Radionuclides
and Heavy Metals in the Environment,
https://doi.org/10.1007/978-3-030-21638-2_2

Table 1 Physical and chemical properties of lead (adopted from Kaye 1995; Lide and Frederikse 1995)

Atomic symbol: Pb
Atomic number: 82
Atomic weight: 207.2 a.m.u
Density: 11.342 g/cm^3 ($\times 1000$ for kg/m^3)
State: solid
Ionizing potential: 7.417 eV
Melting point: 327.46 °C
Boiling point: 1749 °C
Specific heat: 0.129 J/g.K ($\times 1000$ for J/g.K)
Heat of fusion: 4.799 kJ mol^{-1}
Heat of vaporization: 177.7 kJ mol^{-1}
Element abundance
Solar system: 1.03×10^{-8}%
Earth's surface: 14 mg kg^{-1}
Earths ocean: 3×10^{-5} mg l^{-1}

earth's crust (Table 1) and is spread to the atmosphere and environment by dust storms, volcanic movements and water passing through the rocks.

In this section the focus is on the radioactive ^{210}Pb, which is difficult to study chemically and physically, depending on the relatively low radioactive energy. Pb has 49 radioisotopes of which only three are stable (Table 2).

2 Radioactive Lead

Naturally occurring radioactive lead nuclei have ^{214}Pb ($t_{1/2} = 26.8$ min) from ^{238}U, ^{210}Pb ($t_{1/2} = 22.3$ years) from ^{238}U, ^{211}Pb ($t_{1/2} = 36.1$ min) from ^{235}U, and ^{212}Pb ($t_{1/2} = 10.64$ h) from ^{232}Th. Due to its relatively long half-life, ^{210}Pb is the most remarkable among other radioisotopes. Although the half-lives of other radionuclides are relatively small in the geological time scale, they are continuously reproduced, because they are members of the U and Th disintegration series and degraded by these core nuclei. Different fractions of radioactive lead are found in significant amounts in nature (Valkovic 2000). When ^{210}Pb breaks down to decay product ^{210}Bi ($t_{1/2} = 5.013$ days), high energy Bremsstrahlung causes beta radiation, which is a major problem for detector systems. High-energy beta radiation increases background radiation, so it is necessary to take care when using lead shields in detectors. If the detection systems are protected from external background radiation then they can produce strong Bremsstrahlung radiation, which causes the spectroscopic systems to take incorrect measurements.

Table 2 List of lead nuclides (adopted from Kaye 1995; Lide and Frederikse 1995)

Nuclide	Half-Life	Abundance
Pb-181	45 ms	
Pb-182	55 ms	
Pb-183	300 ms	
Pb-184	0.55 s	
Pb-185	4.1 s	
Pb-186	4.83 s	
Pb-187	18.3 s	
Pb-188	24 s	
Pb-189	51 s	
Pb-190	1.2 m	
Pb-191	1.33 m	
Pb-192	3.5 m	
Pb-193	~2 m	
Pb-194	12.0 m	
Pb-195	~15 m	
Pb-196	37 m	
Pb-197	8 m	
Pb-198	2.40 h	
Pb-199	90 m	
Pb-200	21.5 h	
Pb-201	9.33 h	
Pb-202	5.25E4 years	
Pb-203	51.873 h	
Pb-204	>1.4E17 years	1.400%
Pb-205	1.53E+7 years	
Pb-206	Stable	24.100%
Pb-207	Stable	22.100%
Pb-208	Stable	52.400%
Pb-209	3.253 h	
Pb-210	22.3 years	
Pb-211	36.1 m	
Pb-212	10.64 h	
Pb-213	10.2 m	
Pb-214	26.8 m	
Pb-215	36 s	
Pb-187m	15.2 s	
Pb-191m	2.18 m	
Pb-193m	5.8 m	
Pb-195m	15.0 m	
Pb-197m	43 m	
Pb-199m	12.2 m	
Pb-201m	61 s	

(continued)

Table 2 (continued)

Nuclide	Half-Life	Abundance
Pb-202m	3.53 h	
Pb-203m	6.3 s	
Pb-204m	67.2 m	
Pb-205m	5.54 ms	
Pb-207m	0.805 s	
Pb-193m2	5.8 m	
Pb-203m2	0.48 s	

^{210}Pb emits 16.96 keV of 84% intensity and 63.50 keV of beta radiation with intensity of 16%. When detecting ^{210}Pb, high-purity Ge detectors detect 46.54 keV gamma emission energies at the intensity of 4.25%.

3   Modelling of ^{210}Pb

Spatial and spatiotemporal modelling techniques were developed in order to clearly determine the distribution of ^{210}Pb, which is the most effective environmental element of Pb in terms of radioactivity (Külahcı and Şen 2009). The determination of the risk of the activity after determining the coordinates of the sampling points is important for the environmental distribution of the studied variable. In one operational example, ^{210}Pb embodiment the probability distribution function (PDF) for 44 stations in the Hazar Lake in Turkey as in Fig. 1a and the spatial distribution in Fig. 1b. In this study, the Spatiotemporal Point Cumulative Semivariogram (STPCSV) method proposed by Külahcı and Şen (2009) models are employed for the aquatic distribution of ^{210}Pb both spatially and temporally.

Equation 1 is the basis of the STPCSV method.

$$\gamma(d_i) = \frac{1}{2} \sum_{j=1}^{n} \left[(CVt)_j - (CVt)_{j-1} \right]^2 \tag{1}$$

where $\gamma(d_i)$ is the semivariogram value at distance d; C_j corresponds to the concentration in station j; V is the speed of surface water and finally t is the desired time for the ^{210}Pb exchange in future. This simple equation does what many mixed equations cannot do. With the help of the STPCSV graphs (Külahcı and Şen 2009), the effect radius (range) of each station for ^{210}Pb can be calculated and they can be adapted to all micro and macro systems. Figure 2 indicates the temporal change of ^{210}Pb depending on the radius of action (Külahcı and Şen 2009).

A Monte Carlo Simulation for 44 stations is applied with each 44 lead concentration values, which were taken from the surface water. Since, the standard errors in the concentration variation are statistically acceptable, an approximation can be made by assuming that the data have relatively homogeneous change. Prior to the

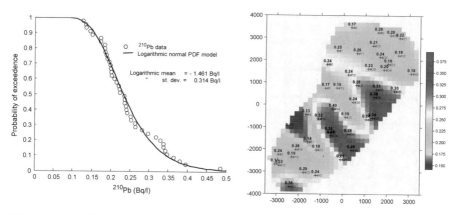

Fig. 1 (**a**) Logarithmic normal PDF model for ^{210}Pb data, (**b**) ^{210}Pb concentration map (Bq l^{-1}) (Adopted from Külahcı and Şen 2009)

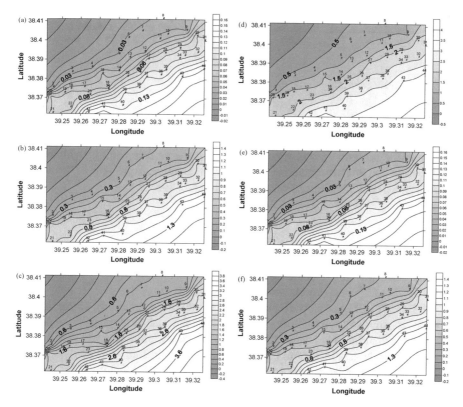

Fig. 2 Simulation curves calculated with STPCSV for Hazar Lake, Turkey. (**a**) Radius of influence for 1 km at 1 h; (**b**) radius of influence for 1 km at 3 h; (**c**) radius of influence for 1 km at 5 h; (**d**) radius of influence for 3 km at 5 h; (**e**) radius of influence for 3 km at 1 h; (**f**) radius of influence for 3 km at 3 h

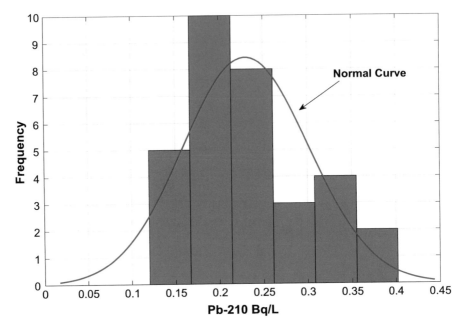

Fig. 3 ²¹⁰Pb histogram plot

simulation analysis, performance of some statistical calculations in a gradual manner helps to evaluate and interpret the results. Figure 3 exposes the statistical distribution in the form of ²¹⁰Pb histogram that shows the corresponding environmental distribution and this histogram indicates the mathematical expression of the normal PDF is given as, PDF (Feller 1968; Patel and Read 1996).

$$y = f\left(x|\mu,\sigma\right) = \frac{1}{\sigma\sqrt{2\pi}} e^{\frac{-(x-\mu)^2}{2\sigma^2}} \tag{2}$$

The corresponding cumulative probability distribution (CDF) (Papoulis and Pillai 2002) of ²¹⁰Pb, is presented in Fig. 4.

The graph in Fig. 4 corresponds to the cumulative probability change and it shows the cumulative probability (Feller 2008) in the desired range with the parameters of the distribution given in Table 3. The necessary inferences and useful interpretations can be obtained from Fig. 4 with reflections in Table 4.

In this table X indicates the entries of the vector in at X = field; $F(X)$ is the corresponding values of the CDF at the entries of X; LB (UB) is the lower (upper) bounds for the confidence interval.

Figure 5 shows a PDF that matches the histogram for Pb data. The main window displays data sets using a probability histogram, in which the height of each rectangle is the fraction of data points that lie in the bin divided by the width of the bin (MathWorks 2019). The sum of the fields of rectangles is 1.

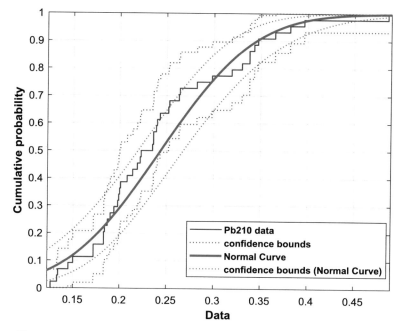

Fig. 4 ^{210}Pb Cumulative probability exchange

Table 3 ^{210}Pb CDF plot parameters

Distribution:	Normal	
Log likelihood:	49.5841	
Mean:	0.244	
Variance:	0.006	
Parameter	*Estimate*	*Std. error*
μ	0.244	0.012
σ	0.080	0.008

Table 4 Cumulative probability prediction of ^{210}Pb

	Normal curve		
X	F(X)	Lower bounds	Upper bounds
0.1000	0.0350	0.0108	0.0924
0.1400	0.0956	0.0434	0.1837
0.1800	0.2111	0.1263	0.3222
0.2200	0.3827	0.2741	0.5015
0.2600	0.5816	0.4631	0.6931
0.3000	0.7613	0.6475	0.8514
0.3400	0.8878	0.7946	0.9460
0.3800	0.9572	0.8942	0.9857
0.4200	0.9869	0.9522	0.9973
0.4600	0.9968	0.9811	0.9996
0.5000	0.9994	0.9935	1.0000

Fig. 5 Probability density function of ^{210}Pb

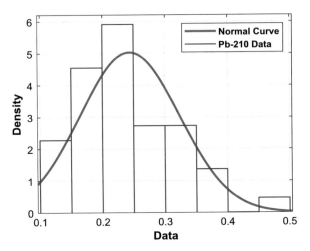

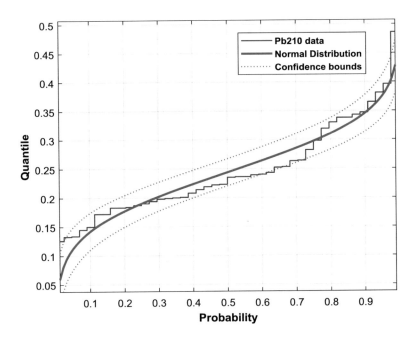

Fig. 6 Quantile (inverse CDF) plot of ^{210}Pb

The percentage point function (PPF) of the ^{210}Pb is calculated as in the inverse cumulative distribution function (inverse CDF) in Fig. 6, which helps PDF calculation. Based on this PDF the PPF and the corresponding X value are calculated for the CDF leading to results in Table 5.

Table 5 Percent point function results

Probability	Normal distribution		
	F/P	LB	UB
0.01	0.059174	0.013488	0.10486
0.108	0.145543	0.114173	0.176914
0.206	0.178601	0.151392	0.20581
0.304	0.202982	0.178006	0.227959
0.402	0.223978	0.200176	0.247781
0.5	0.243659	0.220227	0.267091
0.598	0.26334	0.239537	0.287142
0.696	0.284336	0.259359	0.309312
0.794	0.308717	0.281508	0.335926
0.892	0.341775	0.310404	0.373146
0.99	0.428144	0.382458	0.47383

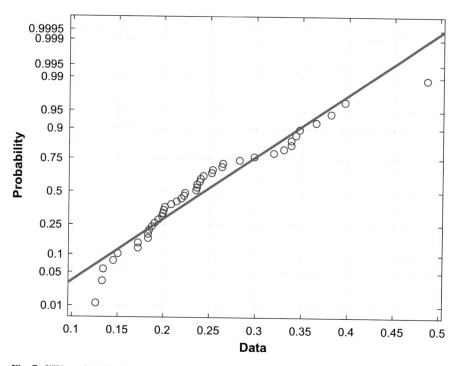

Fig. 7 ^{210}Pb probability change

The probability changes of the data are given in Fig. 7 and the probability change calculations are given in Table 6.

The survivor function performs a kind of reliability analysis of the data at hand. This function is the probability of a variable greater than X (Fig. 8).

Table 6 Probability change results of ^{210}Pb

	Normal distribution		
X	$F(X)$	LB	UB
0.1	0.035029	0.010811	0.092398
0.14	0.095583	0.043379	0.183656
0.18	0.211063	0.126321	0.322212
0.22	0.382722	0.274071	0.501542
0.26	0.581627	0.463092	0.693137
0.3	0.761289	0.647514	0.851363
0.34	0.887789	0.794581	0.946005
0.38	0.957216	0.894215	0.98571
0.42	0.986914	0.952161	0.997291
0.46	0.996814	0.981107	0.999636
0.5	0.999386	0.993512	0.999966

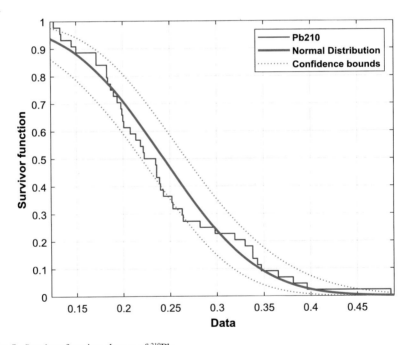

Fig. 8 Survivor function change of ^{210}Pb

The ratio of the PDF to the survival function gives the hazard change in ^{210}Pb (Table 7). The integral of this function leads to the cumulative hazard function in Fig. 9. The cumulative hazard calculation for each X value results as in Table 8. Calculations are made by consideration of 95% confidence limits. The environmental impacts of ^{210}Pb can be calculated effectively in this way (Patel and Read 1996; Spiegel et al. 1982).

Table 7 Survivor function results of ^{210}Pb

| X | Normal distribution | | |
	S(X)	LB	UB
0.1	0.964971	0.907602	0.989189
0.14	0.904417	0.816344	0.956621
0.18	0.788937	0.677788	0.873679
0.22	0.617278	0.498458	0.725929
0.26	0.418373	0.306863	0.536908
0.3	0.238711	0.148637	0.352486
0.34	0.112211	0.053995	0.205419
0.38	0.042784	0.01429	0.105785
0.42	0.013086	0.002709	0.047839
0.46	0.003186	0.000364	0.018893
0.5	0.000614	3.44E-05	0.006488

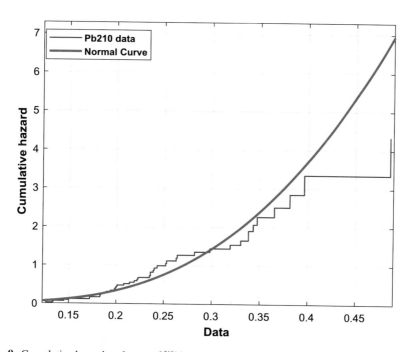

Fig. 9 Cumulative hazard exchange of ^{210}Pb

4   Monte Carlo Simulation of ^{210}Pb Data

Monte Carlo simulation is a technique that looks for the answer to the question of how a model responds to randomly generated inputs or events. Generally, it can be summarized in three steps:

Table 8 Cumulative hazard function results of ^{210}Pb

	Normal distribution		
X	Hazard (X)	LB	UB
0.1	0.0357	0.0969	0.0109
0.14	0.1005	0.2029	0.0443
0.18	0.2371	0.3889	0.1350
0.22	0.4824	0.6962	0.3203
0.26	0.8714	1.1814	0.6219
0.3	1.4325	1.9062	1.0427
0.34	2.1874	2.9189	1.5827
0.38	3.1516	4.2482	2.2464
0.42	4.3362	5.9113	3.0399
0.46	5.7491	7.9188	3.9690
0.5	7.396	10.2777	5.0378

1. Randomly formed, N input or event,
2. Creation of a simulation for each N event,
3. Collection and evaluation the outputs from the simulation.

In this evaluation, the average output value and the output values distribution are obtained with the minimum and maximum output values. Monte Carlo uses probability theory to solve a physical and or mathematical event. The system performs the characterization of the entered data and simulates as much as desired outputs corresponding to the desired time frame. For example, if N different probability outputs are desired then the same numbers of output values are generated according to the probability theory. Finally, the simulation results average corresponds with the observations. The Monte Carlo method is a stochastic model for simulation of reality and it helps to prepare sample experiments. Such simulations are used in the study of system outputs with a large number of variables in stochastic structure (Brémaud 2013; Rubinstein and Kroese 2016).

5   Monte Carlo Forecasting

Monte Carlo simulation can predict future operation possibilities and obtain meaningful inferences. Monte Carlo simulations are useful to make estimates following below steps (Mun 2006):

1. A model is obtained for the estimation of series to be used,
2. Observation series, any inference residue and conditional variance are employed from the sample data,
3. Simulation provides several sample paths for on the desired estimation.

To determine the environmental distribution of the lead, in addition to the classical spatial methods, the modelling and simulation techniques here also give meaningful results. In particular, they provide a different perspective for predicting the

Table 9 ARIMA model of ^{210}Pb

ARIMA (14, 1, 1) model (Gaussian distribution):				
	Values	Standard error	T-statistic	P-value
Constant	0.0036192	0.0016141	2.2422	0.024949
AR(1)	−0.18177	0.24868	−0.73095	0.46481
AR(2)	−0.026117	0.24989	−0.10452	0.91676
AR(3)	−0.003746	0.21262	−0.017619	0.98594
AR(4)	−0.1012	0.24321	−0.41611	0.67733
AR(5)	−0.002084	0.26057	−0.007998	0.99362
AR(6)	−0.33684	0.25186	−1.3374	0.18109
AR(7)	−0.25006	0.22902	−1.0918	0.2749
AR(8)	−0.087773	0.22811	−0.38478	0.7004
AR(9)	0.13004	0.23581	0.55145	0.58132
AR(10)	−0.022881	0.22438	−0.10197	0.91878
AR(11)	−0.40474	0.25434	−1.5913	0.11153
AR(12)	−0.097401	0.29713	−0.32781	0.74306
AR(13)	−0.25968	0.20738	−1.2522	0.21049
AR(14)	−0.10775	0.26893	−0.40065	0.68868
MA(1)	−1	0.19191	−5.2108	1.880e−07
Variance	0.0043306	0.00149	2.9009	0.00372

relevant variable prospectively. Monte Carlo estimation can achieve not only point estimations or a standard error, but also a full distribution for future events. Monte Carlo Simulation estimations are by statistics, through an Auto Regressive Integrated Moving Average (ARIMA) and probability of occurrence (P) event after the standard deviations and variances calculations of proposed different elements (e.g., in here, ^{210}Pb concentrations). Table 9 gives the ARIMA model for the available data according to the normal (Gaussian) PDF. The answer to the question "What would be the concentration values under the same conditions after the 44th cycle of ^{210}Pb?" can be searched easily by the Monte Carlo technique. Figure 10 yields such estimations with 95% confidence level. In order to estimate the concentration of the raw ^{210}Pb data after a further 44 cycles, 100 different pathways are identified with the help of computer, and hence, the simulation of the paths is obtained. Here, the cycle represents the measurement of a new concentration value of ^{40}Pb. After 44 real-time measured values, the answer to the question of what will be the result of 40 measurements looking forward. Monte Carlo simulations are real stochastic simulations. They define the final situation based on the initial state and distribution functions. The latter may be static or dynamic. Any process can show a random characteristic for many reasons. The abrupt changes in the estimation values in Fig. 10 vary accordingly. The reasons for these arbitrary characteristics can be listed as follows:

- A real physical process is stochastic as the events are observed at large scales.
- The process is based on large-scale fluctuations such as the random display of the process due to our lack of information.

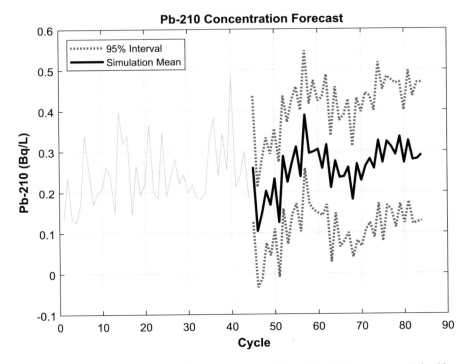

Fig. 10 Monte Carlo Simulation of ^{210}Pb with 95% confidence. The black curve was obtained by simulating the course of 40 cycles of forward time. This curve is the result of the average of 100 different simulation curves

- The process also depends on human behaviour. For example, the form and style of sampling from the research area is important as in this case.
- The model installed for the system may not adequately represent the process of the system.

As a result, if we have something that we cannot know for certain, it shows a stochastic property.

Furthermore, the use of conditional mean and variance models gives meaningful information for variable estimation. In the following example, generalized autoregressive conditional heteroscedastic (GARCH) model is used for the analysis of univariate systems as in ^{210}Pb case.

Herein, GARCH (P,Q) model is

$$\varepsilon_t = \sigma_t z_t \tag{3}$$

and

$$\sigma^2 t = \kappa + \gamma_1 \sigma^2_{t-1} + \ldots + \gamma_P \sigma^2_{t-P} + \alpha_1 \varepsilon^2_{t-1} + \ldots + \alpha Q \varepsilon^2_{t-Q} \tag{4}$$

Table 10 GARCH conditional variance time series model for ^{210}Pb

ARIMA (14, 0, 0) model (*t* distribution):				
	Value	Standard error	*T*-statistic	*P*-value
Constant	3.74E−06	0.22456	1.66E−05	0.9999
AR (14)	1	0.07192	13.852	1.24E−43
DoF	2.4071	1.0119	2.3788	0.01737
GARCH (4, 1) conditional variance model (t distribution):				
	Value	*Standard error*	*T-statistic*	*P-value*
Constant	2.00E−07	0.016981	1.18E−05	0.9999
GARCH (4)	0.0010781	0.00743	0.14511	0.88463
ARCH (1)	0.99892	1.6632	0.6006	0.54811
DoF	2.4071	1.0119	2.3788	0.0137

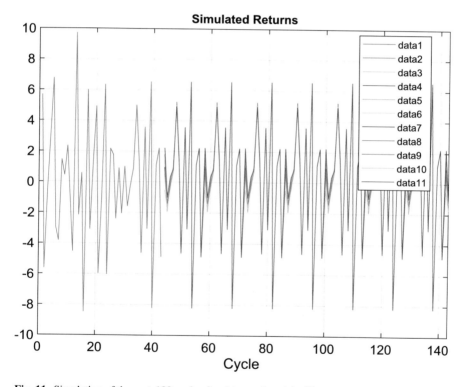

Fig. 11 Simulation of the post-100 cycle after 44 samples of the ^{210}Pb

They are all in the form of Gaussian or Student-*t* PDF innovation (independent). κ is a constant (MathWorks 2019). z_t is degree of freedom (DoF); γ and α are GARCH and autoregressive conditional heteroscedastic (ARCH) coefficients, respectively. The parameters of the GARCH model are *P* and *Q*. The GARCH polynomial consists of lagged conditional variances and is represented by *P*, which is

(a)

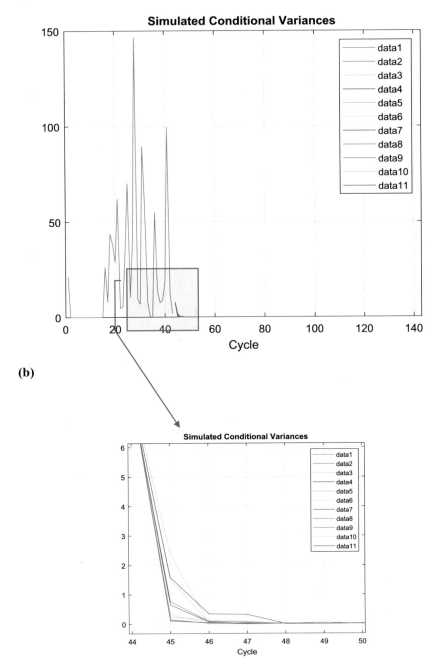

Fig. 12 (**a, b**) Simulated conditional variances of ^{210}Pb

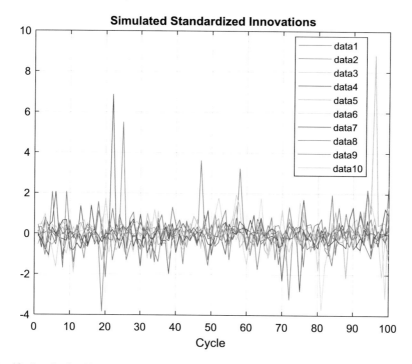

Fig. 13 Standardized innovations for estimation of ^{210}Pb

the number of GARCH terms. An ARCH polynomial is composed of lagged squared innovations and is represented by Q as the number of ARCH terms. Table 10 gives GARCH and ARCH calculations, which have been computed by means of MATLAB, Python, R or SAS software. In this study MATLAB programming language is used.

Innovations are standardized using the square root of conditional variance operations, and the standardized innovations for estimation are obtained as in Fig. 13. The results in Table 10 help to get simulations. Monte Carlo method generates 10 sample paths for the returns, innovations, and conditional variances during 100-period future horizon by the usages of the observed returns, inferred residuals and conditional variances as pre-sample data (MathWorks 2019; Palmer et al. 1990).

The simulation results in Fig. 11 show rather stable volatility over the forecast horizon.

The decrease of volatility in the simulation returns is due to smaller conditional variances over the forecast horizon (see Fig. 12a, b).

Innovations are standardized by using the square root of the conditional variance process and they plotted on the forecast horizon in Fig. 13.

The fitted model assumes that the standardized innovations follow a standard Student's t PDF. Thus, the simulated innovations have larger values than would be expected from a Gaussian innovation PDF.

References

Brémaud P (2013) Markov chains: Gibbs fields, Monte Carlo simulation, and queues. Springer, New York

Feller W (1968) Probability theory and its applications. John Wiley & Sons, New York

Feller W (2008) An introduction to probability theory and its applications. John Wiley & Sons, New York

Kaye G (1995) Kaye & Laby-tables of physical and chemical constants. IOP Publishing Ltd, Bristol, U.K.

Külahcı F, Şen Z (2009) Spatio-temporal modeling of ^{210}Pb transportation in lake environments. J Hazard Mater 165:525–532

Lide DR, Frederikse HPR (1995) Handbook of chemistry and physics 1913–1995. CRC Press, Boca Raton, FL

MathWorks I (2019) Distribution fitting. https://uk.mathworks.com/help/stats/model-data-using-the-distribution-fitting-tool.html. Accessed Feb 2019

Mun J (2006) Modeling risk: applying Monte Carlo simulation, real options analysis, forecasting, and optimization techniques. John Wiley & Sons, New York

Palmer T, Mureau R, Molteni F (1990) The monte carlo forecast. Weather 45:198–207

Papoulis A, Pillai SU (2002) Probability, random variables, and stochastic processes. Tata McGraws-Hill International Edition, New York

Patel JK, Read CB (1996) Handbook of the normal distribution. CRC Press, New York

Rubinstein RY, Kroese DP (2016) Simulation and the Monte Carlo method. John Wiley & Sons, New York

Spiegel MR, Schiller J, Srinivasan RA (1982) Probability and statistics. McGraws-Hill International Edition, New York

Valkovic V (2000) Radioactivity in the Environment: Physicochemical aspects and applications. Elsevier, Amsterderm, The Netherlands

Lead Pollution and Human Exposure: Forewarned is Forearmed, and the Question Now Becomes How to Respond to the Threat!

Natasha, Camille Dumat, Muhammad Shahid, Sana Khalid, and Behzad Murtaza

Abstract At the global scale, persistent lead (Pb) pollution has been considered as a major threat for human health due to its exposure through numerous pathways. The scientific literature regarding potential adverse health effects of Pb is mainly related with human exposure through ingestion, occupational exposure, or dermal absorption. Lead exposure mainly occurs when Pb dust or fumes are inhaled, or when Pb is ingested via contaminated hands, food, water, cigarette, or clothing. Lead entering the digestive and respiratory system is released to the blood and distributed to the whole body. However, data regarding exposomics of Pb (all possible human exposures to Pb including diet, lifestyle, and endogenous sources) is much more limited. Most research on the health effects of Pb focused on the individual routes, mostly soil and water ingestion pathway.

In this chapter, we summarize the current scientific information regarding human exposure to Pb through the various possible routes, their bioaccessible fraction and its accumulation in different body organs. Moreover, the contamination of different exposure media, Pb concentration and the associated human health indices have been calculated to assess the possible carcinogenic and non-carcinogenic risk. These studies indicate that Pb exposure may cause potential human health hazards which suggested being the cause of various organ disorders. Thus, the possible

Natasha · M. Shahid · S. Khalid · B. Murtaza
Department of Environmental Sciences, COMSATS University Islamabad,
Islamabad, Pakistan

C. Dumat (✉)
Centre d'Etude et de Recherche Travail Organisation Pouvoir (CERTOP), UMR5044,
Université de Toulouse—Jean Jaurès, Toulouse Cedex 9, France

Université de Toulouse, INP-ENSAT, Avenue de l'Agrobiopole, Auzeville-Tolosane, France

Association Réseau-Agriville, Castanet-Tolosan, France
e-mail: camille.dumat@ensat.fr

© Springer Nature Switzerland AG 2020
D. K. Gupta et al. (eds.), *Lead in Plants and the Environment*, Radionuclides
and Heavy Metals in the Environment,
https://doi.org/10.1007/978-3-030-21638-2_3

exposure to potentially toxic element Pb through occupational exposure (in addition to workers members of the general public may also be affected due to lead containing releases), atmospheric exposure, skin-contact, water, soil and plant ingestion have been described in four tables and four figures. Based on this current state of knowledge, strategies can then be proposed to reduce human Pb exposure in a context of increasing human density in highly anthropogenic urban areas, and ultimately other metals that have been less studied.

Keywords Lead · Environmental contamination · Human exposure · Toxicity · Health risk

1 Introduction

Among common environmental toxic substances that negatively affect living organisms, lead (Pb) is the most commonly occurring and toxic element (Shahid et al. 2011). Actually, after arsenic, Pb is considered as the most hazardous pollutant by the Agency of Toxic Substance and Disease Registry (ATSDR 2015). It is the 37th most abundant element in Earth's crust with its background concentration of about 0.003 ppb in seawater (Sarkar et al. 2011). Lead has been extensively used in paints, building materials, batteries, ammunition, and radiation protection, and for several decades it was also present in car gasoline with extensive diffusion into the atmosphere (Fig. 1).

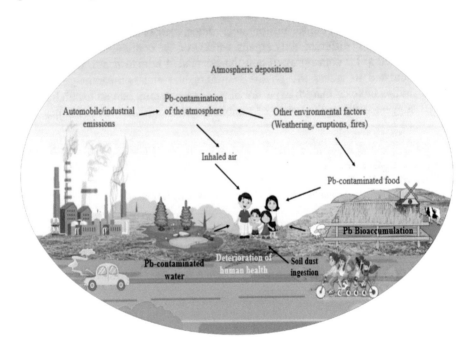

Fig. 1 Possible human exposure to Pb in the environment

Lead is the most toxic cumulative environmental pollutant due to their high mobility and toxicity in soil–plant–human system because of their persistence in the environment that affects multiple body systems and is particularly harmful to young children. Although Pb is not known to have any role in biological systems, it can impact the central nervous system especially in children leading to reduced growth of the brain. These are ranked number two of all hazardous substances by the Agency for Toxic Substances and Disease Registry (ATSDR 2007, 2015).

Human exposure to Pb is mainly through ingestion and inhalation pathway. Acute Pb poisoning may occurs form inhalation of Pb particles/fumes and form ingestion of Pb-contaminated soil or Pb-based paints. After inhalation/ingestion, Pb is transferred to the blood stream and transported to the whole body via circulatory system. The absorption of Pb also depends on the type of Pb inorganic/organic and its bioaccessibility in the gastrointestinal tract. Chronic poisoning from Pb occurs when Pb accumulates in the body over time in soft or mineralizing tissues. This chapter highlights the current status of Pb contamination of the environment and all the possible human exposure routes. Moreover, health risk indices have also been calculated to determine the possible human risk. Finally, few advices are proposed to reduce human exposure to lead and promote human health particularly in high population density urban areas at the global scale (Bories et al. 2018).

2   Uses and Sources of Lead

Lead is the 37th most abundant element of the Earth crust. Naturally, it accounts for high crustal abundance of 14 ppm and (Sarkar et al. 2011). Lead is the most commonly occurring and toxic heavy metal (Shahid et al. 2011). It can be found in all components of the environment (rocks, sediments, soil, air, water or plants) at low concentration; however, it can occur at high concentrations over large areas because of natural and human inputs.

Naturally, Pb is introduced into the environment by volcanic eruptions, wild fires, and various rock/soil erosional processes (Michalak 2001). Moreover, anthropogenically, high levels of Pb are introduced into the environment, particularly due to mining and smelting of the metallic ores. Previously, Pb was used extensively in paints, plastic pipes, building materials, batteries, ammunition, and radiation protection. Elevated levels of Pb have been released into the environment since modern industrialization (Klaminder et al. 2008). Lead has been used as a main component of many products, including plastics, pipes, paints, ceramic ware, cosmetics, wine, glazes and finishes, glassware and gasoline for many years and hence widely distributed and mobilized in the environment.

3 Exposomics of Pb

The exposome of a substance can be defined as the estimation of all the possible exposures of an individual throughout the life to that toxicant (including diet, lifestyle, and endogenous sources) that are not genetic and how those exposures relate to health (Fig. 2). These exposures might begin before birth (at embryonic stage) and includes all environmental and occupational sources.

Lead is considered as highly poisonous element since the time of ancient Greece. Much of the human exposure to Pb comes from anthropogenic activities such as the use of fossil fuels, previously used leaded gasoline, industrial fallouts, and the use of Pb-based paint in homes (Pourrut et al. 2011; Xiong et al. 2016). The primary human exposure to Pb is at homes. As, Pb, and its compounds have been used in a wide variety of products found in and around our homes, including paint, ceramics, pipes and plumbing materials, gasoline, ammunition, Pb-acid batteries (Kawamura and Yanagihara 1998; Lasheen et al. 2008; Johnson et al. 2009; Srivastava et al. 2015). Moreover, occupation exposure of industrial workers to Pb continues to be a matter of health concern worldwide (Mansouri et al. 2018).

Routes of Pb poisoning and its effects, however, are still being discovered. Lead is considered as the most toxic pollutant by the Agency of Toxic Substance and Disease Registry (ATSDR 2015). They are several modes of human exposure to Pb. People can become exposed to Pb through occupational and environmental sources. These may include inhalation of Pb particles from the atmosphere, ingestion of Pb-contaminated soil dust, drinking of Pb-contaminated water and consumption of Pb-contaminated vegetables/crops. Inhalation and ingestion are the two main routes of human exposure to Pb. The all possible routes of human Pb exposure have been presented in Fig. 2. Besides the common exposure pathways, humans can be exposed to Pb through other means (e.g., jewelry, cosmetics, intravenous exposure,

Fig. 2 Different exposure sources of Pb

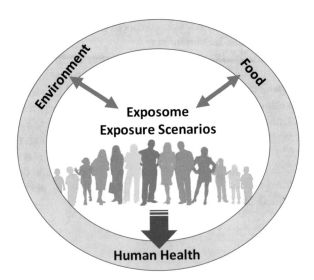

transplacental exposure). That is why in Europe the REACH (Registration, Evaluation, Authorization and Restriction of Chemicals) regulation aims to reduce Pb concentration (and other toxic chemical substances) in the objects currently used by citizens for various activities.

4 Pharmacokinetics of Lead

Pharmacokinetics refers to the pathway of Pb entrance in the body and blood. The pharmacokinetics of Pb in humans is complex. The absorbed Pb (that is not excreted) is traded primarily among the following three compartments: blood, soft tissue (liver, kidneys, lungs, brain, spleen, muscles, and heart) and mineralizing tissues (bones and teeth) (Holstege et al. 2013). The deposition of Pb in these tissues can cause serious health disorders. Several pathways are involved for the enhanced Pb levels in the human body are discussed as follows. In addition to the characterization of sources and transfers, the exposure scenarios, the human practices need also to be studied by social sciences investigations in order to access the sanitary risks.

4.1 Ingestion Pathways

4.2 Soil

Lead, a Ubiquitous and Versatile Element, has been known since ancient times and is freely present in the soil with the background concentration of 27 mg kg^{-1} in soil (Kabata-Pendias 2011). High levels of Pb in soil has been reported in some anthropogenically contaminated sites such as Smelting site (14–7100 mg g^{-1}) (Chlopecka et al. 1996), road dust (105.6 mg kg^{-1}) (Bi et al. 2018), mining site (133.4–45015.97 mg kg^{-1}) (Higueras et al. 2017), industrial contaminated site (42.4–130.7 mg kg^{-1}) (He et al. 2017), former garden soil (1025 mg kg^{-1}) (Egendorf et al. 2018), flooded soil (104–113 mg kg^{-1}) (Antić-Mladenović et al. 2017), shooting ranges (33,000 mg kg^{-1}) (Mariussen et al. 2018) and electronic waste dust (1630–131,000 mg kg^{-1}) (Fujimori et al. 2018). Lead is rarely present in its free form, while mostly found in mineral forms such as galena (PbS), cerussite (PbCO$_3$), anglesite (PbSO$_4$), and minum (Pb$_3$O$_4$) (Shahid 2017).

The potential human exposure to Pb-contaminated soil might result various health hazards. The ingestion of the Pb-contaminated soil is a direct route of human exposure to Pb. Soil ingestion is referred to the consumption of soil which can be attributed to, but not limited to, mouthing, contacting dirty hands, eating dropped food, inhalation of soil dust or consuming soil directly, especially children due to their hand-to-mouth activities (Watt et al. 1993). The densely populated cities, agricultural area, playgrounds, parks are the main sites where people are more vulnerable to Pb poisoning via soil ingestion. Many studies have focused on human disabilities/disorders after Pb exposure related to direct contact with the soil.

Aelion et al. (2013) assessed the soil Pb concentrations in rural and urban areas of South Carolina USA and correlate it with the intellectual disabilities in populations of minority and low-income individuals and children (≤ 6 years of age). The study showed a significant positive association between the mean estimated Pb concentrations and the individuals and children in both urban ($r = 0.38$, $p = 0.0007$) and rural ($r = 0.53$, $p = 0.04$) areas. It was anticipated that the soil metal concentrations contribute greatly to making a population at higher risk for adverse health effects, which is of particular concern for infants and children living in both urban and rural areas.

Moreover, the toxicity of Pb depends on the bioaccessible fractions of Pb in the stomach portion not the total concentration of Pb. It has been noted that the bioaccessible fraction of Pb was lower than the total concentration of Pb in gastric fluid. Only 49% of the total Pb was orally bioaccessible in the surface soils of an urban park of Xiamen City, China (Luo et al. 2012). Bradham et al. (2017) studied the relationship between soil total or bioaccessible Pb and children's blood Pb levels in an urban area in Philadelphia. The bioaccessible concentration of Pb in soil was measured using an in vitro bioaccessibility assay. The observed total soil Pb concentration was 58–2821 mg kg^{-1}; the bioaccessible Pb concentration was 47–2567 mg kg^{-1}. In Children's blood the Pb ranged from 3 to 98 µg L^{-1}. It was estimated that total Pb soil concentration accounts for 23% variability in child blood Pb levels while bioaccessible soil Pb concentration accounted for 26% of variability. Increased level of Pb in blood contribute greatly to the health hazards of the children (Mielke et al. 2017).

Many studies have focused on the health risk from Pb-contaminated soil in terms of Pb bioaccessibility (Table 1) (Lu et al. 2011; Mombo et al. 2016; Attanayake et al. 2017; Mielke et al. 2017; Hu et al. 2018). The health risk related to direct contact with soil (dermal) or through soil ingestion has been considerably reported in literature. The high carcinogenic risk (6.2% higher than the target value) has been reported by Luo et al. (2012) via dermal contact and oral ingestion of the soil. The overall risk was higher for the oral ingestion than dermal contact from Pb-contaminated soil. Soil ingestion has been reported as a direct exposure in children with elevated blood Pb levels (Johnson and Bretsch 2002; Morrison et al. 2013; Han et al. 2018).

Beside the direct human exposure to soil, there are some indirect pathways as well by which human are susceptible towards metal exposure. These pathways include the soil–plant–human pathway. The contamination of soil from various pollution sources (including agriculture, pesticide application, traffic, mining etc.) result in metal buildup in soil and transfer towards food crops. Plants have the ability to store metals in their tissues severalfold higher than the permissible limits (Liu et al. 2013; Popova 2016) and consequently result in food chain contamination and deterioration of the whole ecosystem. The soil–plant–human mobility of the toxic elements has been of great concern from the recent past. Moreover, a set of studies has focused on the health risk assessment of metals by the consumption of contaminated food crops (Zheng et al. 2007; Khan et al. 2008; Maleki and Zarasvand 2008;

Table 1 Bioaccessibility of Pb fractions in human body from different contaminated media

Study area	Pb concentration in the medium	Bioaccessible fractions of Pb	Pb concentration in blood (µg L⁻¹)	References
San Luis Potosi, Mexico	Soil 4062 mg kg⁻¹	Intestine 53%	110	Carrizales et al. (2006)
Liverpool, UK	Dust 646 mg kg⁻¹	Lungs 8.8%	–	Dean et al. (2017)
Karachi, Pakistan	Dust 91 mg kg⁻¹	–	216	Hozhabri et al. (2004)
China	Dust 208 mg kg⁻¹	53%		Han et al. (2012)
USA	Soil 58–2821 mg kg⁻¹	47–2567 mg kg⁻¹	3–98	Bradham et al. (2017)
Brazil	Air 16.65–40.31 µg m⁻³	–	148.9	Peixe et al. (2014)
Italy	Air 45.74 µg m⁻³	–	485	Rapisarda et al. (2016)
Portugal	Soil 6–441 mg kg⁻¹	G 6–260 mg kg⁻¹; GI 0.4–77 mg kg⁻¹	–	Reis et al. (2014)
UK	Dust 452–2435 mg kg⁻¹	Lungs 1.2–8.8%	–	Dean et al. (2017)
China	Dust 220–6348 mg kg⁻¹	Stomach 17.6–76.1%; Intestine 1.2–21.8%	–	
USA	Dust 508.6 mg kg⁻¹	–	50.7	Lanphear et al. (1998)
Idaho	Soil 1686 mg kg; Dust 996	Soil 33% Dust 28%	500–930	von Lindern et al. (2016)
Canada	Soil 74.7	–	14.1	Safruk et al. (2017)
Australia	Soil 2450 mg kg⁻¹	–	>100	Taylor et al. (2014)
France	PM 61.2%	G 25%; GI 15%	–	Uzu et al. (2011)
Africa	Lettuce shoot 77 mg kg⁻¹	49.5%	–	Uzu et al. (2014)
Australia	Soil 1049 mg kg⁻¹	G 100%; GI 2%	–	Xia (2016)
China	Air 180 ng/m³	59.6%	–	Hu et al. (2012)
China	Air 88.2 ng/m³	Lung 59–79%	–	Li et al. (2016)
China	Dust 0.2 mg g⁻¹	G 60%		Hu et al. (2018)

G gastric, *GI* gastrointestinal. WHO limit value of Pb in blood: 10 µg dL⁻¹ or 100 µg L⁻¹, limit bioaccessible fractions <10%.

Khalid et al. 2017; Egendorf et al. 2018; Ndong et al. 2018). Additionally, wastewater irrigation has been of great concern due to the Pb accumulation in the soil and ultimately transferred to the plants (Khalid et al. 2017, 2018).

4.3 Soil Dust

Lead is a potent neurotoxin in children, particularly for toddlers whose brains are developing and who often are exposed to Pb through hand-to-mouth transfer. The ingestion of soil dust is an important Pb exposure pathway by hand-to-mouth transfer or unintentionally by eating food dropped on the floor (Levin et al. 2008; Bierkens et al. 2011; Wang et al. 2018). The major exposure of soil dust, primarily to children, is mainly from playground, urban parks, schoolyards and household dust (Acosta et al. 2009; Reis et al. 2014; Taylor et al. 2014). Soils of playgrounds, day-care centers, kindergartens, schools and sport facility areas in particular are unique because children play on the ground and tend to mouth objects or hands more than adults. Thus, children may ingest significant quantities of Pb-contaminated soil dust, causing severe health disorders.

Besides the direct inhalation/ingestion of soil dust, the dermal absorption of soil dust particles is another main route of exposure to Pb (Benhaddya et al. 2016). Bi et al. (2015) determined the Pb concentration in the urban dust samples showing the range of 79–1544 mg kg^{-1} Pb. Furthermore, the bioaccessible concentration of Pb was measured in stomach and intestinal phase. About 39 and 8.5% of the total Pb in dust was found in stomach and intestinal phase, in vitro. The transfer of Pb via dust ingestion is the main factor contributing to the bioaccessible fractions of the total Pb exposure of children/adult. The concentration of Pb in soil dust (mg kg^{-1}) has been reported in different studies; 99.13 (Abdollahi et al. 2018), 288 (Benhaddya et al. 2016), 300.9 (Christoforidis and Stamatis 2009), 213 (Dehghani et al. 2017), 90.7 (Eqani et al. 2016), 110 (Gope et al. 2017), 110,000 (Leung et al. 2008), 166 (Liu et al. 2014), 7.38 (Solgi and Konani 2016), 120.7 (Suryawanshi et al. 2016), 57.9 (Yoshinaga et al. 2014), 85.3 (Yu et al. 2016) and 4489 (Zheng et al. 2013).

Likewise, Fujimori et al. (2018) measured bioaccessible fractions of Pb in gastrointestinal fluids (in vitro) containing Pb-contaminated dust from electronic waste recycling sites to evaluate the risk of Pb ingestion. The bioaccessible fractions of Pb in gastric fluid were low although the total Pb content was much higher. However, relative high bio-accessibility of organic species of Pb was found in the stomach. Some studies also indicated that Pb citrate present in the soil can increase the Pb bioaccessibility in the stomach (Rasmussen et al. 2011). Health risk from Pb depend on the bioaccessible fractions of Pb in the stomach (Roussel et al. 2010). Many studies have focused on determining the bioaccessible fractions of Pb in intestinal or gastric phase after ingestion Pb-contaminated soil dust (Table 1). Urban agriculture matures in popularity, in post-industrial cities vacant parcels have become a main target for the cultivation (Fig. 3). Many efforts by organizations and individuals to transform the vacant land into fertile, productive agricultural land are

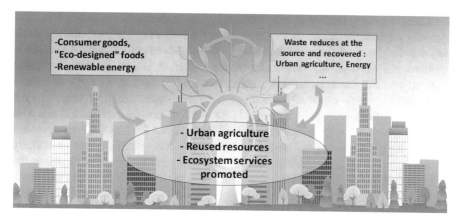

Fig. 3 Recycle and reuse of resources with urban agriculture

spearheaded keen to provide nutritious and healthy fresh produce near and for residents of low-income areas where healthy and fresh food production are limited and/or non-existent because of low purchasing power and market logic of the supermarket location (Gottlieb and Joshi 2010; Alkon and Agyeman 2011). Urban agriculture shows a vital role in food justice movement thus giving fresh produce and range of social benefits. Contamination of Pb in soil is most important and prominent in low income societies where urban agriculture is most common practice (Bernard and McGeehin 2003; Campanella and Mielke 2008; Filippelli and Laidlaw 2010). Urban agriculture and increasing urban food production has been related wide range of benefits. According to public health perspective, urban agriculture has been linked to higher levels of physical activity, improved nutrition, benefits linked with exposure to nature and the increased food security (Davis et al. 2011; Carney et al. 2012; Alaimo et al. 2015; Smith et al. 2018). In order to promote sustainable urban agriculture with low human exposure to metals, it is the quite important to study the biogeochemical cycles of pollutants.

4.4 Exposure Through Pb-Contaminated Water Consumption

Drinking water has been considered as a significant contributor to the individual's overall Pb exposure trajectory with an estimate of 1–20% of the total Pb exposure (Jarvis et al. 2018). Studies have shown that high concentration of Pb in water is linked to the elevated levels of Pb in blood (Akers et al. 2015). World Health Organization has set a provisional guideline of 10 µg L^{-1} of Pb in drinking water (WHO 2004, 2011), while the Pb contamination level goal is set 0 µg L^{-1} due to the recognition of severe Pb toxicity at any exposure level (EPA 2018). It was estimated that about 50% of the world's population rely on groundwater as a drinking source (UNESCO 2015).

There are couple of sources that cause Pb-contamination of groundwater, geo-genic or might be anthropogenic. The water exposure of Pb is an increasingly press-ing matter. The condition of Pb exposure is severe where groundwater is frequently pumped out and used for household chores (Akers et al. 2015). Nowadays, it is noticed that Pb can enter drinking water through the corrosion of Pb-containing pipes (PVC pipes), especially when the chlorine concentration of water is high (Stets et al. 2018). Moreover, brass or chrome-plated brass faucets and fixtures with Pb solder, also add significant amounts of Pb in water. Soldered connections can release high concentration of Pb in drinking water to cause severe toxicity (Tchounwou et al. 2012). More than 36 billion dollars are spent annually in the USA to control corrosion and Pb content in tap water (NWQMC 2017). The release from industries, urban runoff, atmospheric deposition are also the major contribu-tors of Pb in surface water. According to EPA's Toxic Release Inventory, about 4000 facilities released almost 42,581 pounds of Pb and 218,510 pounds of Pb com-pounds in water in 2007 (TRI 2009).

A set of data presents the Pb concentration in drinking water higher than the threshold level ($10 \, \mu g \, L^{-1}$). Deshommes et al. (2016) reported the maximum con-centration of $13200 \, \mu g \, L^{-1}$ of Pb in the 78,791 water samples collected from four Canadian provinces which represent a significant health risk to the consumers. Luu et al. (2009) showed that about 86% of the total water samples from Kandal Province, Cambodia exceed the WHO limit value of Pb in drinking water. Khan et al. (2013) investigated the Pb concentration in drinking water sources (surface/groundwater) collected from Swat valley, Khyber Pakhtunkhwa, Pakistan. All the groundwater samples showed Pb concentration $>10 \, \mu g \, L^{-1}$ due to the geological processes, agricultural and industrial activities, and corrosion of plumbing systems in the proximity of different groundwater sources. Another, possible mode of groundwater Pb contamination is from landfill leachate (Mor et al. 2006; Kubare et al. 2010; Longe and Balogun 2010). The Pb contamination of beds and sediments are due to both natural and anthropogenic inputs (Nazeer et al. 2014). The concen-tration of Pb in drinking water has been reported in Table 2.

4.5   Ingestion of Pb-Contaminated Vegetable/Crops

Mostly plants growing on metal-contaminated sites can accumulate high enough range of metals to cause serious health risk in the consumers. Compared to inhala-tion or dermal contact, ingestion of Pb-contaminated food is reported to be the main pathway (almost 90%) of human exposure (Wang et al. 2011; Xiong et al. 2014b). Consumption of contaminated plants is the primary route of human exposure to Pb. Lead has been widely documented to be taken up by plants and accumulate in the tissues. Many studies have been focused on the uptake and accumulation of Pb in plants through root (Cai et al. 2017; Kumar et al. 2017; Kutrowska et al. 2017) as well as through foliar organs (Uzu et al. 2010; Schreck et al. 2012a; Xiong et al. 2014a; Amato-Lourenco et al. 2016).

Table 2 Concentration of Pb (μg L^{-1}) in drinking water and the estimated health risk . HQ and CR are respectively the Hazard Quotient and Carcinogenic risk

Area	Pb concentration	HQ	CR	References
Tamatave	44	0.314	0.00267	Akers et al. (2015)
India	287	2.050	0.01743	Buragohain et al. (2010)
Saudi Arabia	4.2	0.030	0.00026	Chowdhury et al. (2018)
Canada	13200	94.286	0.80143	Deshommes et al. (2016)
Iran	30.38	0.217	0.00184	Fakhri et al. (2018)
Pakistan	42	0.300	0.00255	Khan et al. (2013)
India	3.19	0.023	0.00019	Kumar et al. (2016)
Cambodia	60	0.429	0.00364	Luu et al. (2009)
Pakistan	24	0.171	0.00146	Muhammad et al. (2011)
Pakistan	1030	7.357	0.06254	Nazeer et al. (2014)
Nigeria	4.49	0.032	0.00027	Odukoya et al. (2017)
Nigeria	3	0.021	0.00018	Omo-Irabor et al. (2008)
Maryland	1.08	0.008	0.00007	Ryan et al. (2000)
Pakistan	146	1.043	0.00886	Ul-Haq et al. (2011)
Thailand	66.9	0.478	0.00406	Wongsasuluk et al. (2014)
China	0.45	0.003	0.00003	Xiao et al. (2019)
Mean	*934.2*	*6.7*	*0.0567*	–
Minimum	*0.5*	*0.003*	*0.0000*	–
Maximum	*13200.0*	*94.3*	*0.8014*	–
S.D.	*820.2*	*5.9*	*0.0498*	–

The uptake of Pb by plants depends on Pb behavior in the soil, in the context of its speciation, solubility, mobility, and bioavailability (Pourrut et al. 2011). The plants growing naturally/artificially on contaminated soil accumulate high concentrations of Pb in their tissues. Leafy vegetables grown in urban/industrial vicinity are more susceptible to accumulate high concentration of Pb in their tissues. Accumulation of Pb by leafy vegetables has been extensively reported (Schreck et al. 2012a, 2014; Mohammad et al. 2017; Bi et al. 2018; Naser et al. 2018; Paltseva et al. 2018).

Crops/vegetables are the major constituent of the human diet, and intake of heavy metal contaminated vegetables can induce numerous human health risks (Shaheen et al. 2016; Rehman et al. 2017; Ghasemidehkordi et al. 2018; Ratul et al. 2018). Thus, vegetable/crop contaminating with Pb can be a potential threat to environmental quality and human health. Many plants are also used for the phytoremediation of Pb-contaminated soil as a low-cost approach for the cleanup of soil (Alaboudi et al. 2018), but the use/dumping of these plants can further effect the environmental quality. Alexander et al. (2006) evaluated the consumption pathway of vegetables grown on Pb-contaminated soil. High accumulation of Pb was observed in spinach, onion and lettuce, widely used as human food, and might cause serious health issues. However, no health risk of consuming vegetables grown in Pb-contaminated soil (Pb-rich mineralization) was observed by Augustsson et al. (2018).

Recent practices are focusing on the Pb accumulation in plants by untreated wastewater irrigation (Jamali et al. 2007; Jan et al. 2010; Khan et al. 2013; Mahmood and Malik 2014; Khalid et al. 2017). Recently, Balkhair and Ashraf (2016) reported that Pb contamination of the vegetables was primarily due to the irrigated with sewage/wastewater. The significant accumulation of Pb in edible portions of the vegetables can cause health disorders on consumption with the estimated health risk index 8.1 and target hazard quotient of 1.8. The evaluation of health risk parameters helps better understand the health hazards from the ingestion of Pb via food products (Chaoua et al. 2018).

Lead accumulation via foliar organs (leaves) has been a matter of concern from the past few years. The increasing demand of local food has forced the farmers to cultivate crops in urban area; along roadsides—heavy traffic emissions, industrial vicinity, along railway lines, and so on. The emitted toxic metals are potentially adsorbed by the aerial organs of the plants and translocated to the whole plant (Shahid et al. 2017). Previously, Schreck et al. (2012a) reported 108, 107, 99, 122 and 171 mg kg^{-1} of Pb accumulation in lettuce shoot after foliar exposure to industrial atmospheric fallouts of a secondary Pb-recycling plant, respectively for 1, 2, 3, 4, and 6 weeks. In another study, Schreck et al. (2012b) evaluated that the Pb content reached about 100 mg kg^{-1} DW in lettuce and parsley and 300 mg kg^{-1} DW in ryegrass after exposure to atmospheric fallouts. Under foliar application of Pb-PMs, about 485 and 214 mg kg^{-1} of Pb was reported, respectively in spinach and cabbage (Xiong et al. 2014a). Similarly, Uzu et al. (2010) showed that after 43 days of Pb-PMs exposure, the thoroughly washed leaves of lettuce contained 335 ± 50 mg kg^{-1} of Pb in shoot tissues. The atmospheric contamination and foliar deposition of Pb can seriously affect vegetable growth and can induce human health risks due to consumption of metal-enriched vegetables. The accumulation of Pb in the edible vegetable/crops from different mode of contamination has been listed in Table 3. It is suggested that particular attention should be given to Pb levels in food stuff for quality assurance and protecting human health injures.

5   Inhalation Pathways

5.1   Occupational Exposure

As lead is widely used in different industries, it is considered as a potent environmental toxin due to its persistent, nonbiodegradable, and toxic nature. Lead poisoning associated with occupational exposure was first reported in 370 BC (Kazantzis 1989). Nowadays, it is common among industrial workers from the past two centuries, when workers were exposed to Pb in smelting, plumbing, printing, painting, and many other industrial activities (Tong et al. 2000). Occupational exposure of workers to Pb could result in elevated levels of Pb in blood and body fluids (Barbosa Jr et al. 2005; Tutuarima 2018). The concentration of Pb in blood should not exceed

Table 3 Lead concentration (mg kg⁻¹) in edible plants and the assessment of risk parameters

Plant	Pb Concentration	HQ	CR	ADD	References
Spinacia oleracea	1.89	0.06	0.000002	0.00023	Alexander et al. (2006)
Spinacia oleracea	0.18	0.01	0.000000	0.00002	Amari et al. (2017)
Abelmoschus esculentus	0.88	0.03	0.000001	0.00011	Balkhair and Ashraf (2016)
Water spinach	0.88	0.03	0.000001	0.00011	Bi et al. (2018)
Triticum æstivum	38.075	1.16	0.000039	0.00464	Chaoua et al. (2018)
Allium cepa	39.3	1.20	0.000041	0.00479	Cherfi et al. (2014)
Lactuca sativa	0.0037	0.00	0.000000	0.00000	Dala-Paula et al. (2018)
Abelmoschus esculentus	10.7	0.33	0.000011	0.00130	Demirezen and Aksoy (2006)
Chili pepper	0.52	0.02	0.000001	0.00006	Fu and Ma (2013)
Solanum nigrum L.	12	0.37	0.000012	0.00146	Ghaderian and Ravandi (2012)
Allium ampeloprasum L.	61	1.86	0.000063	0.00744	Hesami et al. (2018)
Spinacia oleracea	0.05	0.00	0.000000	0.00001	Huang et al. (2014)
Lactuca sativa	1.5	0.05	0.000002	0.00018	Li et al. (2018)
Brassica oleracea	2.86	0.09	0.000003	0.00035	Mahmood and Malik (2014)
Sesamum indicum L.	0.98	0.03	0.000001	0.00012	Mehmood et al. (2018)
Amaranthus dubius	2.38	0.07	0.000002	0.00029	Nabulo et al. (2010)
Spinacia oleracea	1.03	0.03	0.000001	0.00013	Pandey et al. (2012)
Solanum lycopersicum	0.8	0.02	0.000001	0.00010	Rehman et al. (2017)
Tagetes minuta L.	157.6	4.80	0.000163	0.01921	Salazar and Pignata (2014)
Phyla nodiflora	73	2.22	0.000076	0.00890	Yoon et al. (2006)
Dioscorea alata	0.38	0.01	0.000000	0.00005	Zhuang et al. (2009)
Mean	*19.3*	*0.6*	*0.000020*	*0.0024*	–
Minimum	*0.0*	*0.0*	*0.000000*	*0.000000*	–
Maximum	*157.6*	*4.8*	*0.000163*	*0.0192*	–
S.D.	*8.4*	*0.3*	*0.000009*	*0.0010*	–

10 or 100 µg L⁻¹ (WHO 2009). The higher concentration of Pb in the blood can cause significant toxicity, leading to organ and tissues damage vital for human life. Lead in blood have strong affinity for proteins and can dismutase other essential ions in the macromolecules (Chwalba et al. 2018). It may be absorbed from the gastrointestinal tract (via ingestion of Pb-contaminated media) or through the respiratory system (inhalation in Pb-contaminated atmosphere). Once Pb enters the body fluid, it is transported to the whole tissues and body organs by circulation (Li et al. 2004; Andreassen and Rees 2005; Olojo et al. 2005) and consequently severe toxicity to nervous system (Finkelstein et al. 1998; Cecil et al. 2008;), renal failure

(Evans et al. 2017), anemia (Haslam 2003) and immunological disorders (Fenga et al. 2017). The effect of Pb on brain cause reduction of N-acetyl-aspartate, disturbance of the blood brain barrier and reduction in phosphocreatine ratios in frontal lobe gray matter (Trope et al. 2001). The effects of Pb in occupationally exposed workers have been highlighted in Table 4.

Occupational workers are most vulnerable to have high levels of Pb in blood due to their exposure to Pb-contaminated atmosphere. Many researchers have reported high Pb concentration in the blood of workers. Recently, Dobrakowski et al. (2016b) investigated the effect of short-term exposure of Pb on blood morphology and the cytokinesis in occupational workers. The studied group was the allowed to Pb-Zn work for the maintenance of blast furnace. The exposure duration was of 36–44 days. The obtained data showed the Pb concentration of 107 and 491 $\mu g\ L^{-1}$ in blood respectively, before and after Pb exposure. The elevated levels of Pb in blood result in decrease in hemoglobin and cytokinesis increase in leukocytes and platelets while erythrocytes remain unaffected. In another study, Dobrakowski et al. (2016a) reported that 359% increase in the blood Pb levels after occupational Pb exposure, which was linked to the decrease in glutathione activity in the blood and increase in oxidative status in the leukocyte cells. The 26%, 42% and 41% decrease in levels of osteopontin, leptin and prolactin respectively was observed in the workers exposed to Pb (Dobrakowski et al. 2017).

Moreover, some studies have focused on determining the transfer and accumulation of Pb in different body parts by quantifying the concentration of Pb in hair, urine, saliva and nails. Gil et al. (2011) reported that Pb concentration was 22.2, 24.3, and 3.03 $\mu g\ L^{-1}$ in urine, hair, and saliva, respectively, with the mean Pb concentration in blood 43.4 $\mu g\ L^{-1}$. Likewise, Molina-Villalba et al. (2015) determined the Pb concentration in urine and hair samples of children near the mining and industrial area. The maximum Pb concentration observed in urine and hair samples was 54.8 and 13.8 $\mu g\ g^{-1}$, respectively. However, gender difference did not show any variation in Pb accumulation. Mirsattari (2001) reported the urine Pb concentration of 697 $\mu g\ L^{-1}$ in service station attendants exposed to tetraethyl lead. The concentration of Pb in blood/urine/nail/saliva of occupationally exposed workers has been highlighted in Table 4.

5.2 Atmospheric Exposure

Presence of heavy metals in the atmosphere has gained considerable attention from the last two decades due to their high capability to deteriorate the atmospheric quality (Pruvot et al. 2006; Douay et al. 2008; Uzu et al. 2011; Schreck et al. 2014) with environmental and sanitary consequences observed. Heavy metals are adsorbed/bound/attached to the particulate matter (PM), also known as ultrafine particles (Schreck et al. 2011), emitted from industrial activities. Beside this, Pb introduction into the atmosphere is mainly due to the metal and ore processing and vehicles using Pb fuel during several decencies, while other sources may include waste

Table 4 Occupational exposure to Pb and its related health effects

Studied site/area	Pb concentration in blood/urine/hair/nail	Effects	References
Battery recycling plant	Blood 441.2 μg L^{-1}	Blood pressure, hypertension	Rapisarda et al. (2016)
Industrial area	Nail 10.57 ppm; Serum 0.08 ppm; Hair 8.08 ppm	–	Mohmand et al. (2015)
Storage battery plant	Blood 248–393 μg L^{-1}; Skeleton 68–74 μg g^{-1}	Skeleton deformities	Ahlgren et al. (1980)
Zinc smelting plant	Blood 19.2 μg 100 mL^{-1}	Chromosome aberrations in lymphocytes	Bauchinger et al. (1976)
Italian glass factory	Blood 400 μg L^{-1}		Carelli et al. (1999)
Zinc-lead works	Blood 370 μg L^{-1}	Alteration in blood morphology, decrease in hematocrit, changes in hematopoietic cytokines	Chwalba et al. (2018)
Zinc-lead works	Blood 491 μg L^{-1}	Decrease in hemoglobin, increase in white blood cells and platelets, decrease level of cytokinesis related to hematopoietic	Dobrakowski et al. (2016a)
Zinc-lead works	Blood 491 μg L^{-1}	Decrease in erythrocyte glutathione, decrease in glutathione-6-phosphate dehydrogenase in erythrocytes and leukocytes, accumulation of lipid peroxidation products	Dobrakowski et al. (2016b)
Glass manufacturing and battery plants	Blood 145–362 μg L^{-1}	End-stage renal disease	Evans et al. (2017)
Iron-steel industry	Blood 43 μg L^{-1}; Urine 22 μg g^{-1} creatinine; Hair 24 μg g^{-1}; Saliva 3 μg L^{-1}	–	Gil et al. (2011)
Industrial works	Blood 96 μg L^{-1}	Blood pressure, hypertension	Han et al. (2018)
Tile industries	Blood 62.9 μg L^{-1}	Disturbance in the levels of antioxidants, increase in CAT, decrease in SOD and GPx, increase in lipid peroxidation	Hormozi et al. (2018)
Acid and alkaline battery production unit	Blood 282–655 μg L^{-1}	Increase in sister chromatid exchanges, DNA fragmentation in lymphocytes, induction of clastogenic and aneugenic effects in peripheral lymphocytes	Palus et al. (2003)

(continued)

Table 4 (continued)

Studied site/area	Pb concentration in blood/urine/hair/ nail	Effects	References
Smelter plant	Blood 2.73 μg L^{-1}	Increase in the number of miscarriages and premature births, adverse changes in the pattern of estrus and menses	Popovic et al. (2005)
Metallurgical, smelter and refinery plant	Blood 267 μg L^{-1}	Increased concentration of cysteine C in serum, cardiovascular diseases	Poręba et al. (2011a)
Industry	Blood 252.8 μg L^{-1}	Risk for left ventricular diastolic function, arterial hypertension	Poręba et al. (2011b)
Mining and metallurgy plant	Blood 37–632 μg L^{-1}	Cardiovascular diseases, increase in L-homoarginine, fibrinogen, C reactive protein and homocysteine, promote atherosclerosis	Prokopowicz et al. (2017)
Shooting range	Blood 182 μg L^{-1}	–	Rocha et al. (2014)
Storage battery plant	Blood 559 μg L^{-1}	Genetic damage in peripheral blood lymphocytes	Vaglenov et al. (2001)
Lead-acid battery manufacturing and recycling plant	Blood 45 μg L^{-1}	Blood pressure, hypertension	Yang et al. (2018)
Industrial area	Serum 0.07 ppm; Hair 8.2 ppm; Nail 18 ppm	–	Mohmand et al. (2015)
Smelter plant	Blood 98 μg L^{-1}	–	Soto-Jimenez and Flegal (2011)

incinerators, emission from Pb-acid batteries manufacturing industries (USEPA 2017a). Lead in the indoor air comes from Pb dust formed by the scrapping and flaking of paints (USEPA 2017b).

Lead-enriched PM has been widely studied previously (Zheng et al. 2004; Schreck et al. 2011; Dewan et al. 2015; Gemeiner et al. 2017). It has been reported that atmospheric PM have high sorption capacity for Pb (Shahid et al. 2017). Many studies have reported the Pb concentration (ng/m^3) associated with PM at different region of the earth; 180.3 (Hu et al. 2012), 209 (Mateos et al. 2018), 136 (Peter et al. 2018), 30–135 (Voutsa and Samara 2002), 118 (Gupta et al. 2007), 40–318 (Dubey et al. 2012), 589 (Piao et al. 2008), 98 (Rizzio et al. 2001), 101 (Quiterio et al. 2004), 121 (Begum et al. 2013), 4400 (von Schneidemesser et al. 2010) and 371 (Talbi et al. 2018). These Pb-PM can be transported over long distances in the troposphere (Oh et al. 2015).

In Europe, the quality standards of most of the heavy metals in the atmosphere have been defined and updated in Directives 2004/107/CE and 2008/50/CE.-According to these Directives, the annual limit value for Pb and in the atmosphere should not be more than 500 ng/m³. The average concentration of Pb in global urban and industrial air is 200–600 ng/m³ and 500–1000 ng/m³ (Shigeta 2000), respectively. The annual mean concentration of Pb in cities ranges between 200 and 600 ng/m³ in Eastern Europe and <100 ng/m³ in Western Europe (Shahid et al. 2017). The mean Pb atmospheric concentration was 50 ng/m³ in Lisbon (Shahid et al. 2017). Moreover, the atmospheric concentration (ng/m³) of elemental Pb has been documented widely in literature; 64 (Bilos et al. 2001), 291 (Momani et al. 2000), 9.2 (Fernández-Olmo et al. 2016), 14.3 (Mohanraj et al. 2004), 210–620 (Vijayanand et al. 2008), 3–137 (Melaku et al. 2008), 2–4700 (Shah et al. 2006), 121 (Hassanvand et al. 2015), 0.83 (Mafuyai et al. 2014) and 299 (Oucher et al. 2015).

Atmospheric PM is considered a significant source of human diseases (Schreck et al. 2012b; Brauer et al. 2015). Although Pb-PM inhalation is a secondary exposure pathway in human, the main route is the ingestion of Pb-contaminated water/soil/plants. Humans directly inhale the Pb-PM from the atmosphere which induces a range of short-term or long-term toxic effects in humans depending on its characteristics (type, size, composition, and persistent nature). Many cardiovascular diseases have been reported that are associated with direct inhalation of Pb enriched PM (Uzu et al. 2011; Kim et al. 2015; Asgary et al. 2017; Kastury et al. 2017). Uzu et al. (2011) reported human Pb toxicity emitting form a Pb recycling plant. The Pb-PM induced significant pro-inflammatory effect in human bronchial epithelial cells. Goix et al. (2014) reported the toxicity of Pb by means of cytotoxicity, human bioaccessibility, and oxidative potential.

Dappe et al. (2018) determined the total amount of Pb-PM that can be inhaled by using inhalation risk assessment models. The abundance of $PbSO_4$ was 42% in $PM_{0.1-1}$ and 10% in PM_{1-10}. When fine particles containing Pb are respired through lungs, consequently transferred to blood via macrophages while coarse particles are expelled out in the digestive tract (Csavina et al. 2014). However, the determination of soluble/bioaccessible fractions of Pb is more important to evaluate health risk than the total concentration. Boisa et al. (2014) measured the bioaccessibility of Pb in PM_{10} by using epithelial lung fluid (in vitro). The % inhalation bioaccessibility measured was from 0.02% to 11%, indicating the rising chance of pulmonary disorders from inhaling Pb-PM. Recently, Dean et al. (2017) determined the bioaccessible fractions of Pb from PM >10 μm fraction in an urban area. The measured inhalable dose of Pb from PM_{10} was 1.3–7.0 ng kg⁻¹ day⁻¹ body weight with a % bioaccessible fraction 1.2–8.8. However, the calculated human risk was low. In contrast, Hu et al. (2012) reported the high carcinogenic risk for both adult and children from the inhalation and ingestion of Pb-enriched-PM. The carcinogenic risk was also observed form the Pb-bound-particles in the filters of room air conditioners due to high contamination of the outer atmosphere (Hu et al. 2018). On another note, the in vivo determination of Pb bioaccessibility may be more accurate than in vitro studies and important step in the determination of direct and accurate human health risk. Moreover, World Health organization has reported the following human diseases/disorders associated with direct inhalation of Pb from the atmosphere; nervous disorders, renal carcinogenicity, cell hyperplasia, cytomegaly and cellular dysplasia, gene mutations, blood pressure, and cardiovascular effects; inhibition of the activity of the cytoplasmic enzyme δ-ALA; and effects on heme biosynthesis (WHO 2000).

5.3 Dermal Absorption

Less data is available of the dermal absorption of Pb for humans, as cutaneous absorption of Pb by humans is very low, <1% (Holstege et al. 2013). Dermal exposure is mostly among people who work with Pb or materials containing Pb (e.g., workplaces, research labs, industries). The amount absorbed through the skin depends on the physical characteristics of the Pb (i.e., organic vs inorganic) and the integrity of the skin. Organic Pb (tetraethyl Pb) is quite easily absorbed by the skin, while inorganic Pb is not absorbed through intact skin (ATSDR 2017).

6 Other Sources of Human Exposure to Lead

6.1 Flaking of the Paints

Homes with Pb-based paints are also considered as a source of human Pb exposure and making occupants at high risk for Pb exposure and eventually Pb poisoning (Body et al. 1991). Children are more prone to ingest Pb dust/particles from chipping paint-making Pb exposure extremely high. United States Environmental Protection Agency have proposed some indoor sources of human Pb exposure, which may contain; old Pb paint on surfaces (walls, windows, frames), flaking of the Pb paint on any surface, tracking Pb contaminated soil/dust from the outdoors into the indoor environment, home repair activities, or even from Pb dust on clothing at occupational site (USEPA 2017c).

Nduka et al. (2008) estimated the extent of use of Pb-based paint from the four cities of Eastern Nigeria. The flaked paint samples were collected from different areas (church, residential, commercial, and schools). The observed concentration of Pb in flaked paints was 39–74 mg kg^{-1}. However, several studies have reported the high concentration of Pb in paints used in buildings, schools, playgrounds (Adebamowo et al. 2007; Turner and Solman 2016; Turner et al. 2016; da Rocha Silva et al. 2018). Human exposure to these elevated Pb levels in paint flakes can cause detrimental health hazards with the high blood Pb concentration (da Rocha Silva et al. 2018).

7 Health Risk Assessment

Although, determining the total and bioaccessible Pb concentration is quite important for the assessment of human Pb exposure, but without estimation human health hazards is of no use. Nowadays, estimating the health risks is getting increased attention worldwide. Many studies have focused on the evaluation of risk

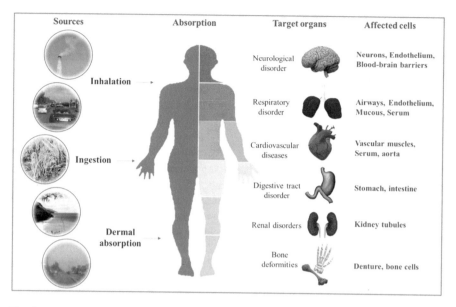

| Sources | Absorption | Target organs | Affected cells |

Fig. 4 Source, absorption and the possible health disorders of environmental exposure to Pb

parameters after human Pb exposure (Zheng et al. 2007; Zhuang et al. 2009; Muhammad et al. 2011; Fu and Ma 2013; Khan et al. 2013; Liu et al. 2013; Zheng et al. 2013; Cherfi et al. 2014; Huang et al. 2014; Mahmood and Malik 2014; Nazeer et al. 2014; Triantafyllidou et al. 2014; Wongsasuluk et al. 2014; Balkhair and Ashraf 2016; Kumar et al. 2016; Odukoya et al. 2017; Rehman et al. 2017; Zhang et al. 2017; Augustsson et al. 2018; Bi et al. 2018; Fakhri et al. 2018; Ji et al. 2018; Li et al. 2018; Sawut et al. 2018; Stets et al. 2018; Xiao et al. 2019). Figure 4 represents the sources and absorption of Pb in human body from different contaminated media, its targeted organs and the possible health disorders.

In this study, we used following parameters to determine the potential health risk; hazard quotient (HQ), cancer risk (CR) and the estimated daily intake/average daily dose (EDI/ADD) for water and plant intake (Tables 2 and 3). The concentration of Pb in edible plants and water was used to measure these parameters. It was observed that HQ values range from 0.003 to 94.3 for water while 0.0001 to 4.8 for plants. The CR assessment showed high carcinogenic risk from Pb-contaminated water with most of the values exceeding the limit (CR > 10^{-4}). However, the CR from plants was lower than the water. This shows that Pb in body can induce both carcinogenic and noncarcinogenic health hazards in human upon exposure. Therefore, Pb transfer and accumulation in human body must be given due consideration to control its health hazards.

8   Conclusions and Perspectives

This bibliographical synthesis perfectly illustrates the long-term common history between the man and the Pb, with still today according to the countries of strong exposures which can occur. In view of the deleterious impacts of Pb, it appears essential to reduce the human exposure to this pollutant and therefore its uses. Based on this current state of knowledge, strategies can then be proposed to reduce human Pb exposure and ultimately other metals that have been less studied.

The first way is to reduce the use of Pb at source by reducing its minimum use for special industrial applications and reducing its concentration. This is the logic of the European REACH regulation: Pb is gradually, replaced in various materials or when it is present in trace for example in sewage sludge, their spreading will depend on the quantities involved to take into accounts the total quantity. The second way is to reduce the diffusion, the transfer of Pb in the environment. This pathway, such as liming soils to reduce the solubility of Pb or using plants that have the ability to phytostabilize Pb by trapping it in their roots, can eventually result in the cultivation of moderately polluted soils by producing healthy foodstuffs. Since Pb is persistent, many urban soils around the world are polluted. The majority of people live in cities today and urban agriculture is growing. It is therefore crucial to develop strategies to reduce pollutant transfers in order to live well even in partially degraded environments at the global scale thanks to the knowledge of biogeochemical mechanisms.

The third way is to restore the quality of the environment and also to better train the populations to the risks related to the use and exposure to chemical substances. This involves regulations that impose management and communication constraints on businesses and the training of users, consumers, and voters on the subject of environmental health and sustainable food. Actually, forewarned is forearmed. It is with this goal of training society on the link between environment and health, that various online educational resources are developed. We can mention the massive open online course "MOOC-TEAM" focus on metals in the environment and human exposure, with accessibility for all the students (Laffont et al. 2018) or the "Réseau-Agriville" pedagogical site dedicated to urban agriculture (https://reseau-agriville.com/, Dumat et al. 2018).

References

Abdollahi S, Karimi A, Madadi M, Ostad-Ali-Askari K, Eslamian S, Singh VP (2018) Lead concentration in dust fall in Zahedan, Sistan and Baluchistan province, Iran. J Geogra Cartog 1. https://doi.org/10.24294/jgc.v1i2.601

Acosta JA, Cano AF, Arocena JM, Debela F, Martínez-Martínez S (2009) Distribution of metals in soil particle size fractions and its implication to risk assessment of playgrounds in Murcia City (Spain). Geoderma 149:101–109

Adebamowo EO, Scott Clark C, Roda S, Agbede OA, Sridhar MKC, Adebamowo CA (2007) Lead content of dried films of domestic paints currently sold in Nigeria. Sci Total Environ 388:116–120

Aelion CM, Davis HT, Lawson AB, Cai B, McDermott S (2013) Associations between soil lead concentrations and populations by race/ethnicity and income-to-poverty ratio in urban and rural areas. Environ Geochem Health 35:1–12

Ahlgren L, Haeger-Aronsen B, Mattsson S, Schütz AJO, Medicine E (1980) In-vivo determination of lead in the skeleton after occupational exposure to lead. Occup Environ Med 37:109–113

Akers DB, MacCarthy MF, Cunningham JA, Annis J, Mihelcic JR (2015) Lead (Pb) contamination of self-supply groundwater systems in coastal Madagascar and predictions of blood lead levels in exposed children. Environ Sci Technol 49:2685–2693

Alaboudi KA, Ahmed B, Brodie GJ (2018) Phytoremediation of Pb and Cd contaminated soils by using sunflower (*Helianthus annuus*) plant. Ann Agricult Sci 63:123–127

Alaimo K, Oleksyk S, Golzynski D (2015) The michigan healthy school action tools generate improvements in school nutrition policies and practices, and student dietary intake. Health Promot Pract 16:401–410

Alexander P, Alloway B, Dourado AJ (2006) Genotypic variations in the accumulation of Cd, Cu, Pb and Zn exhibited by six commonly grown vegetables. Environ Pollut 144:736–745

Alkon AH, Agyeman J (eds) (2011) Cultivating food justice: race, class, and sustainability. MIT Press, p. 404. JSTOR, http://www.jstor.org/stable/j.ctt5vjpc1

Amari T, Ghnaya T, Abdelly C (2017) Nickel, cadmium and lead phytotoxicity and potential of halophytic plants in heavy metal extraction. South Afri J Bot 111:99–110

Amato-Lourenco LF, Moreira TCL, de Oliveira Souza VC, Barbosa F Jr, Saiki M, Saldiva PHN, Mauad TJ (2016) The influence of atmospheric particles on the elemental content of vegetables in urban gardens of Sao Paulo, Brazil. Environ Pollut 216:125–134

Andreassen S, Rees SE (2005) Mathematical models of oxygen and carbon dioxide storage and transport: interstitial fluid and tissue stores and whole-body transport. Crit Rev Biomed Eng 33:265–298

Antić-Mladenović S, Frohne T, Kresović M, Stärk HJ, Tomić Z, Ličina V, Rinklebe J (2017) Biogeochemistry of Ni and Pb in a periodically flooded arable soil: Fractionation and redox-induced (im)mobilization. J Environ Manage 186:141–150

Asgary S, Movahedian A, Keshvari M, Taleghani M, Sahebkar A, Sarrafzadegan N (2017) Serum levels of lead, mercury and cadmium in relation to coronary artery disease in the elderly: a cross-sectional study. Chemosphere 180:540–544

ATSDR (2007) Toxicological profile for arsenic - Agency for toxic substances and disease registry. U.S. Department of Health and Human Services

ATSDR (2015) Agency for toxic substances and disease registry toxicological profile for lead. U.S. Department of Health and Human Services

ATSDR (2017) Environmental health and medicine education. Lead toxicity. What are routes of exposure to lead? U.S. Department of Health and Human Services

Attanayake CP, Hettiarachchi GM, MaQ PGM, Ransom MD (2017) Lead speciation and in vitro bioaccessibility of compost-amended urban garden soils. J Environ Qual 46:1215–1224

Augustsson A, Uddh-Söderberg T, Filipsson M, Helmfrid I, Berglund M, Karlsson H, Hogmalm J, Karlsson A, Alriksson SJ (2018) Challenges in assessing the health risks of consuming vegetables in metal-contaminated environments. Environ Int 113:269–280

Balkhair KS, Ashraf MA (2016) Field accumulation risks of heavy metals in soil and vegetable crop irrigated with sewage water in western region of Saudi Arabia. Saudi J Biol Sci 23:S32–S44

Barbosa F Jr, Tanus-Santos JE, Gerlach RF, Parsons PJ (2005) A critical review of biomarkers used for monitoring human exposure to lead: advantages, limitations, and future needs. Environ Health Perspect 113:1669–1674

Bauchinger M, Schmid E, Einbrodt H, Dresp JJ (1976) Chromosome aberrations in lymphocytes after occupational exposure to lead and cadmium. Mutat Res 40:57–62

Begum BA, Hopke PK, Markwitz A (2013) Air pollution by fine particulate matter in Bangladesh. Atmos Pollut Res 4:75–86

Benhaddya ML, Boukhelkhal A, Halis Y, Hadjel M (2016) Human health risks associated with metals from urban soil and road dust in an oilfield area of Southeastern Algeria. Arch Environ Contam Toxicol 70:556–571

Bernard SM, McGeehin MA (2003) Prevalence of blood lead levels >5µg/dL among US children 1 to 5 years of age and socioeconomic and demographic factors associated with blood of lead levels 5 to 10 µg/dL, Third National Health and Nutrition Examination Survey, 1988–1994. PEDIATRICS 112:6

Bi X, Li Z, Sun G, Liu J, Han Z (2015) In vitro bioaccessibility of lead in surface dust and implications for human exposure: A comparative study between industrial area and urban district. J Hazard Nater 297:191–197

Bi C, Zhou Y, Chen Z, Jia J, Bao XJ (2018) Heavy metals and lead isotopes in soils, road dust and leafy vegetables and health risks via vegetable consumption in the industrial areas of Shanghai, China. Sci Total Environ 619:1349–1357

Bierkens J, Van Holderbeke M, Cornelis C, Torfs R (2011) Exposure through soil and dust ingestion. In: Swatjes FA (ed) Dealing with contaminated sites. Springer, London, New York, pp 261–286

Bilos C, Colombo J, Skorupka C, Presa MR (2001) Sources, distribution and variability of airborne trace metals in La Plata City area, Argentina. Environ Pollut 111:149–158

Body PE, Inglis G, Dolan PR, Mulcahy DE (1991) Environmental lead: a review. Crit Review Environ Cont 20:299–310

Boisa N, Elom N, Dean JR, Deary ME, Bird G, Entwistle JA (2014) Development and application of an inhalation bioaccessibility method (IBM) for lead in the PM10 size fraction of soil. Environ Int 70:132–142

Bories O, Dumat C, Sochacki L, Aubry C (2018) Les agricultures urbaines durables : un vecteur pour la transition écologique. Hors-série 31

Bradham KD, Nelson CM, Kelly J, Pomales A, Scruton K, Dignam T, Misenheimer JC, Li K, Obenour DR, Thomas DJ (2017) Relationship between total and bioaccessible lead on children's blood lead levels in urban residential Philadelphia soils. Environ Sci Technol 51:10005–10011

Brauer M, Freedman G, Frostad J, Van Donkelaar A, Martin RV, Dentener F, Dingenen RV, Estep K, Amini H, Apte JS (2015) Ambient air pollution exposure estimation for the global burden of disease 2013. Environ Sci Technol 50:79–88

Buragohain M, Bhuyan B, Sarma HP (2010) Seasonal variations of lead, arsenic, cadmium and aluminium contamination of groundwater in Dhemaji district, Assam, India. Environ Monit Assess 170:345–351

Cai F, Wu X, Zhang H, Shen X, Zhang M, Chen W, Gao Q, White JC, Tao S, Wang XJN (2017) Impact of TiO2 nanoparticles on lead uptake and bioaccumulation in rice (*Oryza sativa* L.). NanoImpact 5:101–108

Campanella R, Mielke HW (2008) Human geography of New Orleans' urban soil lead contaminated geochemical setting. Environ Geochem Health 30:531–540

Carelli G, Masci O, Altieri A, Castellino N (1999) Occupational exposure to lead—granulometric distribution of airborne lead in relation to risk assessment. Ind Health 37:313–321

Carney PA, Hamada JL, Rdesinski R, Sprager L, Nichols KR, Liu BY, Pelayo J, Sanchez MA, Shannon J (2012) Impact of a community gardening project on vegetable intake, food security and family relationships: a community-based participatory research study. J Community Health 37:874–881

Carrizales L, Razo I, Téllez-Hernández JI, Torres-Nerio R, Torres A, Batres LE, Cubillas AC, Diaz-Barriga F (2006) Exposure to arsenic and lead of children living near a copper-smelter in San Luis Potosi, Mexico: importance of soil contamination for exposure of children. Environ Res 101:1–10

Cecil KM, Brubaker CJ, Adler CM, Dietrich KN, Altaye M, Egelhoff JC, Wessel S, Elangovan I, Hornung R, Jarvis K (2008) Decreased brain volume in adults with childhood lead exposure. PLoS Med 5:e112

Chaoua S, Boussaa S, El Gharmali A, Boumezzough A (2018) Impact of irrigation with wastewater on accumulation of heavy metals in soil and crops in the region of Marrakech in Morocco. J Saudi Soc Agri Sci. https://doi.org/10.1016/j.jssas.2018.02.003

Cherfi A, Abdoun S, Gaci O (2014) Food survey: levels and potential health risks of chromium, lead, zinc and copper content in fruits and vegetables consumed in Algeria. Food Chem Toxicol 70:48–53

Chlopecka A, Bacon J, Wilson M, Kay J (1996) Forms of cadmium, lead, and zinc in contaminated soils from southwest Poland. J Environ Qual 25:69–79

Chowdhury S, Kabir F, Mazumder MAJ, Zahir MH (2018) Modeling lead concentration in drinking water of residential plumbing pipes and hot water tanks. Sci Total Environ 635:35–44

Christoforidis A, Stamatis NJ (2009) Heavy metal contamination in street dust and roadside soil along the major national road in Kavala's region, Greece. Geoderma 151:257–263

Chwalba A, Maksym B, Dobrakowski M, Kasperczyk S, Pawlas N, Birkner E, Kasperczyk A (2018) The effect of occupational chronic lead exposure on the complete blood count and the levels of selected hematopoietic cytokines. Toxicol Appl Pharmacol 355:174–179

Csavina J, Taylor MP, Félix O, Rine KP, Sáez AE, Betterton EA (2014) Size-resolved dust and aerosol contaminants associated with copper and lead smelting emissions: implications for emission management and human health. Sci Total Environ 493:750–756

da Rocha Silva JP, Salles FJ, Leroux IN, da Silva Ferreira APS, da Silva AS, Assunção NA, Nardocci AC, Sayuri Sato AP, Barbosa F Jr, Cardoso MRA, Olympio KPK (2018) High blood lead levels are associated with lead concentrations in households and day care centers attended by Brazilian preschool children. Environ Pollut 239:681–688

Dala-Paula BM, Custódio FB, Knupp EA, Palmieri HE, Silva JBB, Glória MB (2018) Cadmium, copper and lead levels in different cultivars of lettuce and soil from urban agriculture. Environ Pollut 242:383–389

Dappe V, Uzu G, Schreck E, Wu L, Li X, Dumat C, Moreau M, Hanoune B, Ro CU, Sobanska S (2018) Single-particle analysis of industrial emissions brings new insights for health risk assessment of PM. Atmos Pollut Res 9:697–704

Davis C, Curtis C, Levitan RD, Carter JC, Kaplan AS, Kennedy JL (2011) Evidence that 'food addiction' is a valid phenotype of obesity. Appetite 57:711–717

Dean JR, Elom NI, Entwistle JA (2017) Use of simulated epithelial lung fluid in assessing the human health risk of Pb in urban street dust. Sci Total Environ 579:387–395

Dehghani S, Moore F, Keshavarzi B, Beverley AH (2017) Health risk implications of potentially toxic metals in street dust and surface soil of Tehran, Iran. Ecotox Environ Safety 136:92–103

Demirezen D, Aksoy A (2006) Heavy metal levels in vegetables in Turkey are within safe limits for Cu, Zn, Ni and exceeded for Cd and Pb. J Food Qual 29:252–265

Deshommes E, Andrews RC, Gagnon G, McCluskey T, McIlwain B, Doré E, Nour S, Prévost M (2016) Evaluation of exposure to lead from drinking water in large buildings. Water Res 99:46–55

Dewan N, Majestic BJ, Ketterer ME, Miller-Schulze JP, Shafer MM, Schauer JJ, Solomon PA, Artamonova M, Chen BB, Imashev SA (2015) Stable isotopes of lead and strontium as tracers of sources of airborne particulate matter in Kyrgyzstan. Atmos Environ 120:438–446

Dobrakowski M, Pawlas N, Hudziec E, Kozłowska A, Mikołajczyk A, Birkner E, Kasperczyk S (2016a) Glutathione, glutathione-related enzymes, and oxidative stress in individuals with subacute occupational exposure to lead. Environ Toxicol Pharmacol 45:235–240

Dobrakowski M, Boroń M, Czuba ZP, Birkner E, Chwalba A, Hudziec E, Kasperczyk S (2016b) Blood morphology and the levels of selected cytokines related to hematopoiesis in occupational short-term exposure to lead. Toxicol Appl Pharmacol 305:111–117

Dobrakowski M, Kasperczyk A, Czuba Z, Machoń-Grecka A, Szlacheta Z, Kasperczyk S (2017) The influence of chronic and subacute exposure to lead on the levels of prolactin, leptin, osteopontin, and follistatin in humans. Hum Exp Toxicol 36:587–593

Douay F, Pruvot C, Roussel H, Ciesielski H, Fourrier H, Proix N, Waterlot C (2008) Contamination of urban soils in an area of Northern France polluted by dust emissions of two smelters. Water Air Soil Pollut 188:247–260

Dubey B, Pal AK, Singh G (2012) Trace metal composition of airborne particulate matter in the coal mining and non–mining areas of Dhanbad Region, Jharkhand, India. Atmos Pollut Res 3:238–246

Dumat C, Mombo S, Shahid M, Pierart A, Xiong T (2018) "Réseau-Agriville", international network to promote interdisciplinary synergies for Research-Training-Development on sustainable (peri)urban agricultures. Acte du Congrès International Changements et Transitions: enjeux pour les éducations à l'environnement et au développement durable, 7–9 Nov 2017, Toulouse, France

Egendorf SP, Cheng Z, Deeb M, Flores V, Paltseva A, Walsh D, Groffman P, Mielke HW (2018) Constructed soils for mitigating lead (Pb) exposure and promoting urban community gardening: the New York City Clean Soil Bank pilot study. Landscape Urban Plan 175:184–194

EPA (2018) Ground water and drinking water. Basic information about lead in drinking water.

Eqani SAMAS, Kanwal A, Bhowmik AK, Sohail M, Ullah R, Ali SM, Alamdar A, Ali N, Fasola M, Shen HJ (2016) Spatial distribution of dust–bound trace elements in Pakistan and their implications for human exposure. Environ Pollut 213:213–222

Evans M, Discacciati A, Quershi AR, Åkesson A, Elinder CG (2017) End-stage renal disease after occupational lead exposure: 20 years of follow-up. Occup Environ Med 74:396–401

Fakhri Y, Saha N, Ghanbari S, Rasouli M, Miri A, Avazpour M, Rahimizadeh A, Riahi SM, Ghaderpoori M, Keramati H (2018) Carcinogenic and non-carcinogenic health risks of metal (oid) s in tap water from Ilam city, Iran. Food Chem Toxicol 118:204–211

Fenga C, Gangemi S, Di Salvatore V, Falzone L, Libra M (2017) Immunological effects of occupational exposure to lead. Mol Med Rep 15:3355–3360

Fernández-Olmo I, Andecochea C, Ruiz S, Fernández-Ferreras JA, Irabien A (2016) Local source identification of trace metals in urban/industrial mixed land-use areas with daily PM10 limit value exceedances. Atmos Res 171:92–106

Filippelli GM, Laidlaw MAS (2010) The elephant in the playground: confronting lead-contaminated soils as an important source of lead burdens to urban populations. Perspect Biol Med 53:31–45

Finkelstein Y, Markowitz ME, Rosen JF (1998) Low-level lead-induced neurotoxicity in children: an update on central nervous system effects. Brain Res Rev 27:168–176

Fu W, Ma G (2013) The characters and health risk assessment of vegetable Pb in Jilin Suburb. Proc Environ Sci 18:221–226

Fujimori T, Taniguchi M, Agusa T, Shiota K, Takaoka M, Yoshida A, Terazono A, Ballesteros FC, Takigami H (2018) Effect of lead speciation on its oral bioaccessibility in surface dust and soil of electronic-wastes recycling sites. J Hazard Mater 341:365–372

Gemeiner H, de Araujo DT, Sulato ET, Galhardi JA, Gomes ACF, de Almeida E, Menegário AA, Gastmans D, Kiang CH (2017) Elemental and isotopic determination of lead (Pb) in particulate matter in the Brazilian city of Goiânia (GO) using ICP-MS technique. Environ Sci Pollut Res 24:20616–20625

Ghaderian SM, Ravandi AA (2012) Accumulation of copper and other heavy metals by plants growing on Sarcheshmeh copper mining area, Iran. J Geochem Explor 123:25–32

Ghasemidehkordi B, Malekirad AA, Nazem H, Fazilati M, Salavati H, Shariatifar N, Rezaei M, Fakhri Y, Khaneghah AM (2018) Concentration of lead and mercury in collected vegetables and herbs from Markazi province, Iran: a non-carcinogenic risk assessment. Food Chem Toxicol 113:204–210

Gil F, Hernández AF, Márquez C, Femia P, Olmedo P, López-Guarnido O, Pla A (2011) Biomonitorization of cadmium, chromium, manganese, nickel and lead in whole blood, urine, axillary hair and saliva in an occupationally exposed population. Sci Total Environ 409:1172–1180

Goix S, Lévêque T, Xiong TT, Schreck E, Baeza-Squiban A, Geret F, Uzu G, Austruy A, Dumat C (2014) Environmental and health impacts of fine and ultrafine metallic particles: assessment of threat scores. Environ Res 133:185–194

Gope M, Masto RE, George J, Hoque RR, Balachandran S (2017) Bioavailability and health risk of some potentially toxic elements (Cd, Cu, Pb and Zn) in street dust of Asansol, India. Ecotoxicol Environ Saf 138:231–241

Gottlieb R, Joshi A (2010) Food Justice. MIT Press, p. 320

Gupta AK, Karar K, Srivastava A (2007) Chemical mass balance source apportionment of PM10 and TSP in residential and industrial sites of an urban region of Kolkata, India. J Hazard Mater 142:279–287

Han Z, Guo X, Zhang B, Liao J, Nie L (2018) Blood lead levels of children in urban and suburban areas in China (1997–2015): temporal and spatial variations and influencing factors. Sci Total Environ 625:1659–1666

Haslam RH (2003) Lead poisoning. Paediatr Child Health 8:509–510

Hassanvand MS, Naddafi K, Faridi S, Nabizadeh R, Sowlat MH, Momeniha F, Gholampour A, Arhami M, Kashani H, Zare A (2015) Characterization of PAHs and metals in indoor/outdoor PM 10/PM 2.5/PM 1 in a retirement home and a school dormitory. Sci Total Environ 527:100–110

He K, Sun Z, Hu Y, Zeng X, Yu Z, Cheng H (2017) Comparison of soil heavy metal pollution caused by e-waste recycling activities and traditional industrial operations. Environ Sci Pollut Res 24:9387–9398

Hesami R, Salimi A, Ghaderian SM (2018) Lead, zinc, and cadmium uptake, accumulation, and phytoremediation by plants growing around Tang-e Douzan lead–zinc mine, Iran. Environ Sci Pollut Res 25:8701–8714

Higueras P, Esbrí JM, García-Ordiales E, González-Corrochano B, López-Berdonces MA, García-Noguero EM, Alonso-Azcárate J, Martínez-Coronado A (2017) Potentially harmful elements in soils and holm-oak trees (Quercus ilex L.) growing in mining sites at the Valle de Alcudia Pb-Zn district (Spain)–some clues on plant metal uptake. J Geochem Explor 182:166–179

Holstege C, Huff J, Rowden A, O'Malley R (2013) Pathophysiology and etiology of lead toxicity. Retrieved from Medscape. https://emedicine.medscape.com/article/2060369-overview

Hormozi M, Mirzaei R, Nakhaee A, Izadi S, Dehghan Haghighi JJ (2018) The biochemical effects of occupational exposure to lead and cadmium on markers of oxidative stress and antioxidant enzymes activity in the blood of glazers in tile industry. Toxicol Ind Health 34:459–467

Hozhabri S, White F, Rahbar MH, Agboatwalla M, Luby S (2004) Elevated blood lead levels among children living in a fishing community, Karachi, Pakistan. Arch Environ Health Int J 59:37–41

Hu X, Zhang Y, Ding Z, Wang T, Lian H, Sun Y, Wu J (2012) Bioaccessibility and health risk of arsenic and heavy metals (Cd, Co, Cr, Cu, Ni, Pb, Zn and Mn) in TSP and PM2.5 in Nanjing, China. Atmos Environ 57:146–152

Hu X, Xu X, Ding Z, Chen Y, Lian H (2018) In vitro inhalation/ingestion bioaccessibility, health risks, and source appointment of airborne particle-bound elements trapped in room air conditioner filters. Environ Sci Pollut Res 25:26059–26068

Huang Z, Pan XD, Wu PG, Han JL, Chen Q (2014) Heavy metals in vegetables and the health risk to population in Zhejiang, China. Food Control 36:248–252

Jamali M, Kazi T, Arain M, Afridi H, Jalbani N, Memon A (2007) Heavy metal contents of vegetables grown in soil, irrigated with mixtures of wastewater and sewage sludge in Pakistan, using ultrasonic-assisted pseudo-digestion. J Agron Crop Sci 193:218–228

Jan FA, Ishaq M, Khan S, Ihsanullah I, Ahmad I, Shakirullah M (2010) A comparative study of human health risks via consumption of food crops grown on wastewater irrigated soil (Peshawar) and relatively clean water irrigated soil (lower Dir). J Hazard Mater 179:612–621

Jarvis P, Quy K, Macadam J, Edwards M, Smith M (2018) Intake of lead (Pb) from tap water of homes with leaded and low lead plumbing systems. Sci Total Environ 644:1346–1356

Ji Y, Wu P, Zhang J, Zhang J, Zhou Y, Peng Y, Zhang S, Cai G, Gao G (2018) Heavy metal accumulation, risk assessment and integrated biomarker responses of local vegetables: a case study along the Le'an river. Chemosphere 199:361–371

Johnson DL, Bretsch JK (2002) Soil lead and children's blood lead levels in Syracuse, NY, USA. Environ Geochem Health 24:375–385

Johnson S, Saikia N, Sahu MR (2009) Lead in paints. Centre for Science and Environment, and Pollution Monitoring Laboratory, New Delhi, India

Kabata-Pendias A (2011) Trace elements in soils and plants. CRC Press, Boca Raton, FL

Kastury F, Smith E, Juhasz AL (2017) A critical review of approaches and limitations of inhalation bioavailability and bioaccessibility of metal(loid)s from ambient particulate matter or dust. Sci Total Environ 574:1054–1074

Kawamura A, Yanagihara T (1998) State of charge estimation of sealed lead-acid batteries used for electric vehicles. In: Power Electronics Specialists Conference, 1998. PESC 98 Record. 29th Annual IEEE, vol 1. IEEE, p. 583–587

Kazantzis G (1989) Lead: ancient metal-modern menace? In: Smith MA, Grant LD, Sors AI (eds) Lead exposure and child development. Springer, Dordrecht, p. 119–128

Khalid S, Shahid M, Dumat C, Niazi NK, Bibi I, Gul Bakhat HFS, Abbas G, Murtaza B, Javeed HMR (2017) Influence of groundwater and wastewater irrigation on lead accumulation in soil and vegetables: Implications for health risk assessment and phytoremediation. Int J Phytoremediation 19:1037–1046

Khalid S, Shahid M, Natasha Bibi I, Sarwar T, Shah A, Niazi N (2018) A review of environmental contamination and health risk assessment of wastewater use for crop irrigation with a focus on low and high-income countries. Int J Environ Res Public Health 15:895

Khan K, Lu Y, Khan H, Zakir S, Khan S, Khan AA, Wei L, Wang T (2013) Health risks associated with heavy metals in the drinking water of Swat, northern Pakistan. J Environ Sci 25:2003–2013

Khan S, Cao Q, Zheng Y, Huang Y, Zhu Y (2008) Health risks of heavy metals in contaminated soils and food crops irrigated with wastewater in Beijing, China. Environ Pollut 152:686–692

Kim KH, Kabir E, Kabir S (2015) A review on the human health impact of airborne particulate matter. Environ Int 74:136–143

Klaminder J, Bindler R, Rydberg J, Renberg I (2008) Is there a chronological record of atmospheric mercury and lead deposition preserved in the mor layer (O-horizon) of boreal forest soils? Geochim Cosmochim Acta 72:703–712

Kubare M, Mutsvangwa C, Masuku C (2010) Groundwater contamination due to lead (Pb) migrating from Richmond municipal landfill into Matsheumhlope aquifer: evaluation of a model using field observations. Drink Water Eng Sci Discuss 3:251–269

Kumar B, Smita K, Flores LC (2017) Plant mediated detoxification of mercury and lead. Arab J Chem 10:S2335–S2342

Kumar M, Rahman MM, Ramanathan A, Naidu R (2016) Arsenic and other elements in drinking water and dietary components from the middle Gangetic plain of Bihar, India: health risk index. Sci Total Environ 539:125–134

Kutrowska A, Małecka A, Piechalak A, Masiakowski W, Hanć A, Barałkiewicz D, Andrzejewska B, Zbierska J, Tomaszewska BJ (2017) Effects of binary metal combinations on zinc, copper, cadmium and lead uptake and distribution in *Brassica juncea*. J Trace Elem Med Biol Fert Soil 44:32–39

Laffont L, Dumat C, Pape S, Leroy A, Piran K, Bassette C, Altinier A, Jolibois F (2018) MOOC-TEAM "Environmental transfers of metallic contaminants: an inclusive education in environmental health for ecological transition". International CESS Congress. Toulouse (France), 4–6 July 2018

Lanphear BP, Matte TD, Rogers J, Clickner RP, Dietz B, Bornschein RL, Succop P, Mahaffey KR, Dixon S, Galke W (1998) The contribution of lead-contaminated house dust and residential soil to children's blood lead levels: a pooled analysis of 12 epidemiologic studies. Environ Res 79:51–68

Lasheen M, Sharaby C, El-Kholy N, Elsherif I, El-Wakeel S (2008) Factors influencing lead and iron release from some Egyptian drinking water pipes. J Hazard Mater 160:675–680

Leung AO, Duzgoren-Aydin NS, Cheung K, Wong MH (2008) Heavy metals concentrations of surface dust from e-waste recycling and its human health implications in southeast China. Environ Sci Technol 42:2674–2680

Levin R, Brown MJ, Kashtock ME, Jacobs DE, Whelan EA, Rodman J, Schock MR, Padilla A, Sinks T (2008) Lead exposures in US children, 2008: implications for prevention. Environ Health Perspect 116:1285

Li GJ, Zhang LL, Lu L, Wu P, Zheng W (2004) Occupational exposure to welding fume among welders: alterations of manganese, iron, zinc, copper, and lead in body fluids and the oxidative stress status. J Occup Environ Med/Am College Occup Environ Med 46:241

Li SW, Li HB, Luo J, Li HM, Qian X, Liu MM, Bi J, Cui XY, Ma LQ (2016) Influence of pollution control on lead inhalation bioaccessibility in PM2. 5: a case study of 2014 Youth Olympic Games in Nanjing. Environ Int 94:69–75

Li X, Li Z, Lin CJ, Bi X, Liu J, Feng X, Zhang H, Chen J, Wu T (2018) Health risks of heavy metal exposure through vegetable consumption near a large-scale Pb/Zn smelter in central China. Ecotox Environ Safety 161:99–110

Liu E, Yan T, Birch G, Zhu Y (2014) Pollution and health risk of potentially toxic metals in urban road dust in Nanjing, a mega-city of China. Sci Total Environ 476:522–531

Liu X, Song Q, Tang Y, Li W, Xu J, Wu J, Wang F, Brookes PC (2013) Human health risk assessment of heavy metals in soil–vegetable system: a multi-medium analysis. Sci Total Environ 463:530–540

Longe E, Balogun M (2010) Groundwater quality assessment near a municipal landfill, Lagos, Nigeria. Res J Appl Sci Eng Technol 2:39–44

Lu Y, Yin W, Huang L, Zhang G, Zhao Y (2011) Assessment of bioaccessibility and exposure risk of arsenic and lead in urban soils of Guangzhou City, China. Environ Geochem Health 33:93–102

Luo XS, Ding J, Xu B, Wang YJ, Li HB, Yu S (2012) Incorporating bioaccessibility into human health risk assessments of heavy metals in urban park soils. Sci Total Environ 424:88–96

Luu TTG, Sthiannopkao S, Kim KW (2009) Arsenic and other trace elements contamination in groundwater and a risk assessment study for the residents in the Kandal Province of Cambodia. Environ Int 35:455–460

Mafuyai GM, Eneji IS, Sha'Ato R (2014) Concentration of heavy metals in respirable dust in Jos metropolitan area, Nigeria. Open J Air Pollut 3:10

Mahmood A, Malik RN (2014) Human health risk assessment of heavy metals via consumption of contaminated vegetables collected from different irrigation sources in Lahore, Pakistan. Arab J Chem 7:91–99

Maleki A, Zarasvand MA (2008) Heavy metals in selected edible vegetables and estimation of their daily intake in Sanandaj, Iran. Southeast Asian J Trop Med Public Health 39:335

Mansouri MT, Muñoz-Fambuena I, Cauli O (2018) Cognitive impairment associated with chronic lead exposure in adults. Neurol Psych Brain Res 30:5–8

Mariussen E, Johnsen IV, Strømseng AE (2018) Application of sorbents in different soil types from small arms shooting ranges for immobilization of lead (Pb), copper (Cu), zinc (Zn), and antimony (Sb). J Soils Sedim 18:1558–1568

Mateos A, Amarillo A, Carreras H, González C (2018) Land use and air quality in urban environments: human health risk assessment due to inhalation of airborne particles. Environ Res 161:370–380

Mehmood S, Saeed DA, Rizwan M, Khan MN, Aziz O, Bashir S, Ibrahim M, Ditta A, Akmal M, Mumtaz MA (2018) Impact of different amendments on biochemical responses of sesame (Sesamum indicum L.) plants grown in lead-cadmium contaminated soil. Plant Physiol Biochem 132:345–355

Melaku S, Morris V, Raghavan D, Hosten C (2008) Seasonal variation of heavy metals in ambient air and precipitation at a single site in Washington, DC. Environ Pollut 155:88–98

Michalak AM (2001) Feasibility of contaminant source identification for property rights enforcement. In: Anderson TL, Hill PJ (eds) The technology of property rights. Rowman & Littlefield Publishers, Lanham, MD, p. 123–145

Mielke HW, Gonzales CR, Powell ET (2017) Soil lead and children's blood lead disparities in pre- and post-hurricane Katrina New Orleans (USA). Int J Environ Res Public Health 14:407

Mirsattari S (2001) Urine lead levels in service station attendants exposed to tetraethyl lead. J Res Med Sci 6:2

Mohammad I, Daniel A, Kiyawa S, Kutama A (2017) Phyto-accumulation of lead and chromium in common edible green-leafy vegetables consumed in Dutse Metropolis, Jigawa State, Nigeria. Int J Chem Mater Environ Res 4:131–136

Mohanraj R, Azeez PA, Priscilla T (2004) Heavy metals in airborne particulate matter of urban Coimbatore. Arch Environ Contam Toxicol 47:162–167

Mohmand J, Eqani SAMAS, Fasola M, Alamdar A, Mustafa I, Ali N, Liu L, Peng S, Shen H (2015) Human exposure to toxic metals via contaminated dust: Bio-accumulation trends and their potential risk estimation. Chemosphere 132:142–151

Molina-Villalba I, Lacasaña M, Rodríguez-Barranco M, Hernández AF, Gonzalez-Alzaga B, Aguilar-Garduño C, Gil F (2015) Biomonitoring of arsenic, cadmium, lead, manganese and mercury in urine and hair of children living near mining and industrial areas. Chemosphere 124:83–91

Momani KA, Jiries AG, Jaradat QM (2000) Atmospheric deposition of Pb, Zn, Cu, and Cd in Amman, Jordan. Turk J Chem 24:231–238

Mombo S, Foucault Y, Deola F, Gaillard I, Goix S, Shahid M, Schreck E, Pierart A, Dumat C (2016) Management of human health risk in the context of kitchen gardens polluted by lead and cadmium near a lead recycling company. J Soils Sedim 16:1214–1224

Mor S, Ravindra K, Dahiya RP, Chandra A (2006) Leachate characterization and assessment of groundwater pollution near municipal solid waste landfill site. Environ Monit Assess 118:435–456

Morrison D, Lin Q, Wiehe S, Liu G, Rosenman M, Fuller T, Wang J, Filippelli G (2013) Spatial relationships between lead sources and children's blood lead levels in the urban center of Indianapolis (USA). Environ Geochem Health 35:171–183

Muhammad S, Shah MT, Khan S (2011) Health risk assessment of heavy metals and their source apportionment in drinking water of Kohistan region, northern Pakistan. Microchem J 98:334–343

Nabulo G, Young S, Black C (2010) Assessing risk to human health from tropical leafy vegetables grown on contaminated urban soils. Sci Total Environ 408:5338–5351

Naser H, Rahman M, Sultana S, Quddus M, Hossain M (2018) Heavy metal accumulation in leafy vegetables grown in industrial areas under varying levels of pollution. Bangl J Agricult Res 43:39–51

Nazeer S, Hashmi MZ, Malik RN (2014) Heavy metals distribution, risk assessment and water quality characterization by water quality index of the River Soan, Pakistan. Ecol Indic 43:262–270

Ndong M, Mise N, Okunaga M, Kayama F (2018) Cadmium, arsenic and lead accumulation in rice grains produced in Senegal river valley. Fund Toxicol Sci 5:87–91

Nduka J, Orisakwe O, Maduawguna C (2008) Lead levels in paint flakes from buildings in Nigeria: a preliminary study. Toxicol Ind Health 24:539–542

NWQMC (2017) National water quality monitoring council. National Water Monitoring News. https://acwi.gov/monitoring/

Odukoya A, Olobaniyi S, Oluseyi T, Adeyeye U (2017) Health risk associated with some toxic elements in surface water of Ilesha gold mine sites, southwest Nigeria. Environ Nanotech Monit Manag 8:290–296

Oh HR, Ho CH, Kim J, Chen D, Lee S, Choi YS, Chang LS, Song CK (2015) Long-range transport of air pollutants originating in China: a possible major cause of multi-day high-PM10 episodes during cold season in Seoul, Korea. Atmos Environ 109:23–30

Olojo E, Olurin K, Mbaka G, Oluwemimo A (2005) Histopathology of the gill and liver tissues of the African catfish Clarias gariepinus exposed to lead. Afr J Biotechnol 4:117–122

Omo-Irabor OO, Olobaniyi SB, Oduyemi K, Akunna J (2008) Surface and groundwater water quality assessment using multivariate analytical methods: a case study of the Western Niger Delta, Nigeria. Phys Chem Earth 33:666–673

Oucher N, Kerbachi R, Ghezloun A, Merabet H (2015) Magnitude of air pollution by heavy metals associated with aerosols particles in algiers. Energy Procedia 74:51–58

Paltseva A, Cheng Z, Deeb M, Groffman PM, Shaw RK, Maddaloni M (2018) Accumulation of arsenic and lead in garden-grown vegetables: factors and mitigation strategies. Sci Total Environ 640:273–283

Palus J, Rydzynski K, Dziubaltowska E, Wyszynska K, Natarajan A, Nilsson RJ (2003) Genotoxic effects of occupational exposure to lead and cadmium. Mutat Res/Genet Toxicol Environ Mutagen 540:19–28

Pandey R, Singh VP, Pandey SK (2012) Lead (Pb) accumulation study in plants and vegetables cultivates around coal mines and power plant of Singrauli District. Int J Pharma Sci Res 3:5079–5086

Peixe TS, Nascimento ES, Silva CS, Bussacos MA (2014) Occupational exposure profile of Pb, Mn, and Cd in nonferrous Brazilian sanitary alloy foundries. Toxicol Ind Health 30:701–713

Peter AE, Nagendra SS, Nambi IM (2018) Comprehensive analysis of inhalable toxic particulate emissions from an old municipal solid waste dumpsite and neighborhood health risks. Atmos Pollut Res 9:1021–1031

Piao F, Sun X, Liu S, Yamauchi T (2008) Concentrations of toxic heavy metals in ambient particulate matter in an industrial area of northeastern China. Front Med China 2:207–210

Popova E (2016) Accumulation of heavy metals in the "soil-plant" system. In: AIP Conference Proceedings, vol 1772. AIP Publishing, p. 050006

Popovic M, McNeill FE, Chettle DR, Webber CE, Lee CV, Kaye WE (2005) Impact of occupational exposure on lead levels in women. Environ Health Perspect 113:478–484

Poręba R, Gać P, Poręba M, Antonowicz-Juchniewicz J, Andrzejak R (2011a) Relation between occupational exposure to lead, cadmium, arsenic and concentration of cystatin C. Toxicology 283:88–95

Poręba R, Gać P, Poręba M, Antonowicz-Juchniewicz J, Andrzejak R (2011b) Relationship between occupational exposure to lead and local arterial stiffness and left ventricular diastolic function in individuals with arterial hypertension. Toxicol Appl Pharmacol 254:342–348

Pourrut B, Shahid M, Dumat C, Winterton P, Pinelli E (2011) Lead uptake, toxicity, and detoxification in plants. Rev Environ Contam Toxicol 213:113–136

Prokopowicz A, Sobczak A, Szuła-Chraplewska M, Zaciera M, Kurek J, Szołtysek-Bołdys I (2017) Effect of occupational exposure to lead on new risk factors for cardiovascular diseases. Occup Environ Med 74:366–373

Pruvot C, Douay F, Hervé F, Waterlot C (2006) Heavy metals in soil, crops and grass as a source of human exposure in the former mining areas (6 pp). J Soil Sediment 6:215–220

Quiterio SL, Da Silva CRS, Arbilla G, Escaleira V (2004) Metals in airborne particulate matter in the industrial district of Santa Cruz, Rio de Janeiro, in an annual period. Atmos Environ 38:321–331

Rapisarda V, Ledda C, Ferrante M, Fiore M, Cocuzza S, Bracci M, Fenga C (2016) Blood pressure and occupational exposure to noise and lead (Pb) A cross-sectional study. Toxicol Ind Health 32:1729–1736

Rasmussen PE, Beauchemin S, Chénier M, Levesque C, MacLean LC, Marro L, Jones-Otazo H, Petrovic S, McDonald LT, Gardner HD (2011) Canadian house dust study: lead bioaccessibility and speciation. Environ Sci Technol 45:4959–4965

Ratul A, Hassan M, Uddin M, Sultana M, Akbor M, Ahsan M (2018) Potential health risk of heavy metals accumulation in vegetables irrigated with polluted river water. Int Food Res J 25:329–338

Rehman ZU, Khan S, Brusseau ML, Shah MT (2017) Lead and cadmium contamination and exposure risk assessment via consumption of vegetables grown in agricultural soils of five-selected regions of Pakistan. Chemosphere 168:1589–1596

Reis A, Patinha C, Wragg J, Dias A, Cave M, Sousa A, Batista MJ, Prazeres C, Costa C, da Silva EF (2014) Urban geochemistry of lead in gardens, playgrounds and schoolyards of Lisbon, Portugal: assessing exposure and risk to human health. Appl Geochem 44:45–53

Rizzio E, Bergamaschi L, Valcuvia M, Profumo A, Gallorini M (2001) Trace elements determination in lichens and in the airborne particulate matter for the evaluation of the atmospheric pollution in a region of northern Italy. Environ Int 26:543–549

Rocha ED, Sarkis JES, Maria de Fátima HC, dos Santos GV, Canesso C (2014) Occupational exposure to airborne lead in Brazilian police officers. Int J Hyg Environ Health 217:702–704

Roussel H, Waterlot C, Pelfrêne A, Pruvot C, Mazzuca M, Douay F (2010) Cd, Pb and Zn oral bioaccessibility of urban soils contaminated in the past by atmospheric emissions from two lead and zinc smelters. Arch Environ Contam Toxicol 58:945–954

Ryan PB, Huet N, MacIntosh DL (2000) Longitudinal investigation of exposure to arsenic, cadmium, and lead in drinking water. Environ Health Perspect 108:731–735

Safruk AM, McGregor E, Aslund MLW, Cheung PH, Pinsent C, Jackson BJ, Hair AT, Lee M, Sigal EA (2017) The influence of lead content in drinking water, household dust, soil, and paint on blood lead levels of children in Flin Flon, Manitoba and Creighton, Saskatchewan. Sci Total Environ 593:202–210

Salazar MJ, Pignata ML (2014) Lead accumulation in plants grown in polluted soils. Screening of native species for phytoremediation. J Geochem Explor 137:29–36

Sarkar D, Datta R, Hannigan R (2011) Concepts and applications in environmental geochemistry. Elsevier, Amsterderm, The Netherlands

Sawut R, Kasim N, Maihemuti B, Hu L, Abliz A, Abdujappar A, Kurban M (2018) Pollution characteristics and health risk assessment of heavy metals in the vegetable bases of northwest China. Sci Total Environ 642:864–878

Schreck E, Foucault Y, Geret F, Pradère P, Dumat C (2011) Influence of soil ageing on bioavailability and ecotoxicity of lead carried by process waste metallic ultrafine particles. Chemosphere 85:1555–1562

Schreck E, Bonnard R, Laplanche C, Leveque T, Foucault Y, Dumat C (2012a) DECA: a new model for assessing the foliar uptake of atmospheric lead by vegetation, using *Lactuca sativa* as an example. J Environ Manage 112:233–239

Schreck E, Foucault Y, Sarret G, Sobanska S, Cécillon L, Castrec-Rouelle M, Uzu G, Dumat C (2012b) Metal and metalloid foliar uptake by various plant species exposed to atmospheric industrial fallout: mechanisms involved for lead. Sci Total Environ 427:253–262

Schreck E, Dappe V, Sarret G, Sobanska S, Nowak D, Nowak J, Stefaniak EA, Magnin V, Ranieri V, Dumat C (2014) Foliar or root exposures to smelter particles: consequences for lead compartmentalization and speciation in plant leaves. Sci Total Environ 476–477:667–676

Shah MH, Shaheen N, Jaffar M, Khalique A, Tariq SR, Manzoor S (2006) Spatial variations in selected metal contents and particle size distribution in an urban and rural atmosphere of Islamabad, Pakistan. J Environ Manage 78:128–137

Shaheen N, Irfan NM, Khan IN, Islam S, Islam MS, Ahmed MK (2016) Presence of heavy metals in fruits and vegetables: health risk implications in Bangladesh. Chemosphere 152:431–438

Shahid M (2017) Biochemical behaviour of heavy metals in soil-plant system. Higher Education Commission of Pakistan

Shahid M, Pinelli E, Pourrut B, Silvestre J, Dumat C (2011) Lead-induced genotoxicity to Vicia faba L. roots in relation with metal cell uptake and initial speciation. Ecotox Environ Safety 74:78–84

Shahid M, Dumat C, Khalid S, Schreck E, Xiong T, Niazi NK (2017) Foliar heavy metal uptake, toxicity and detoxification in plants: a comparison of foliar and root metal uptake. J Hazard Mater 325:36–58

Shigeta T (2000) Environmental investigation in Pakistan. Pak-EPA/JICA, Islamabad

Smith D, Thompson C, Harland K, Storm P, Shelton N (2018) Identifying populations and areas at greatest risk of household food insecurity in England. Appl Geogr 91:21–31

Solgi E, Konani R (2016) Assessment of lead contamination in soils of urban parks of Khorramabad, Iran. Health Scope 5:e36056

Soto-Jimenez MF, Flegal AR (2011) Childhood lead poisoning from the smelter in Torreón, México. Environ Res 111:590–596

Srivastava D, Singh A, Baunthiyal M (2015) Lead toxicity and tolerance in plants. J Plant Sci Res 2:123

Stets EG, Lee CJ, Lytle DA, Schock MR (2018) Increasing chloride in rivers of the conterminous US and linkages to potential corrosivity and lead action level exceedances in drinking water. Sci Total Environ 613:1498–1509

Suryawanshi P, Rajaram B, Bhanarkar A, Rao CC (2016) Determining heavy metal contamination of road dust in Delhi, India. Atmosfera 29:221–234

Talbi A, Kerchich Y, Kerbachi R, Boughedaoui M (2018) Assessment of annual air pollution levels with PM1, PM2.5, PM10 and associated heavy metals in Algiers, Algeria. Environ Pollut 232:252–263

Taylor MP, Mould SA, Kristensen LJ, Rouillon M (2014) Environmental arsenic, cadmium and lead dust emissions from metal mine operations: Implications for environmental management, monitoring and human health. Environ Res 135:296–303

Tchounwou PB, Yedjou CG, Patlolla AK, Sutton DJ (2012) Heavy metal toxicity and the environment. Molecular, clinical and environmental toxicology. In: Luch A (ed) Molecular, clinical and environmental toxicology. Experientia supplementum, vol 101. Springer, Basel, p. 133–164

Tong S, Schirnding YE, Prapamontol T (2000) Environmental lead exposure: a public health problem of global dimensions. Bull World Health Organ 78:1068–1077

TRI (2009) TRI Explorer Chemical Report. U.S. Environmental Protection Agency. http://www.epa.gov/ triexplorer and select Lead

Triantafyllidou S, Le T, Gallagher D, Edwards M (2014) Reduced risk estimations after remediation of lead (Pb) in drinking water at two US school districts. Sci Total Environ 466:1011–1021

Trope I, Lopez-Villegas D, Cecil KM, Lenkinski RE (2001) Exposure to lead appears to selectively alter metabolism of cortical gray matter. Pediatrics 107:1437–1442

Turner A, Solman KR (2016) Lead in exterior paints from the urban and suburban environs of Plymouth, south west England. Sci Total Environ 547:132–136

Turner A, Kearl ER, Solman KR (2016) Lead and other toxic metals in playground paints from South West England. Sci Total Environ 544:460–466

Tutuarima J (2018) The correlation between lead level and hemoglobin, hematocrit, cystatin C serum, SGOT and SGPT levels of car paint workers. Health Not 2:387–397

Ul-Haq N, Arain M, Badar N, Rasheed M, Haque Z (2011) Drinking water: a major source of lead exposure in Karachi, Pakistan. East Mediterr Health J 17:882–886

UNESCO (2015) Water for a sustainable world. United Nations World Water Assessment Programme. Division of Water Sciences, UNESCO 06134 Colombella, Perugia, Italy

USEPA (2017a) United States, Environmental Protection Agency: Basic Information about Lead Air Pollution

USEPA (2017b) Protect Your Family from Exposures to Lead. United States Environmental Protection Agency

USEPA (2017c) United States Environmental Protection Agency. Protect Your Family from Exposures to Lead

Uzu G, Sobanska S, Sarret G, Munoz M, Dumat C (2010) Foliar lead uptake by lettuce exposed to atmospheric fallouts. Environ Sci Technol 44:1036–1042

Uzu G, Sauvain JJ, Baeza-Squiban A, Riediker M, Sánchez Sandoval Hohl M, Val S, Tack K, Denys S, Pradere P, Dumat C (2011) In vitro assessment of the pulmonary toxicity and gastric availability of lead-rich particles from a lead recycling plant. Environ Sci Technol 45:7888–7895

Uzu G, Schreck E, Xiong T, Macouin M, Lévêque T, Fayomi B, Dumat C (2014) Urban market gardening in Africa: foliar uptake of metal(loid)s and their bioaccessibility in vegetables; implications in terms of health risks. Water Air Soil Pollut 225:2185

Vaglenov A, Creus A, Laltchev S, Petkova V, Pavlova S, Marcos R (2001) Occupational exposure to lead and induction of genetic damage. Environ Health Perspect 109:295–298

Vijayanand C, Rajaguru P, Kalaiselvi K, Selvam KP, Palanivel M (2008) Assessment of heavy metal contents in the ambient air of the Coimbatore city, Tamilnadu, India. J Hazard Mater 160:548–553 von Lindern I, Spalinger S, Stifelman ML, Stanek LW, Bartrem C (2016) Estimating children's soil/dust ingestion rates through retrospective analyses of blood lead biomonitoring from the Bunker Hill Superfund Site in Idaho. Environ Health Perspect 124:1462–1470

von Schneidemesser E, Stone EA, Quraishi TA, Shafer MM, Schauer JJ (2010) Toxic metals in the atmosphere in Lahore, Pakistan. Sci Total Environ 408:1640–1648

Voutsa D, Samara C (2002) Labile and bioaccessible fractions of heavy metals in the airborne particulate matter from urban and industrial areas. Atmos Environ 36:3583–3590

Wang B, Lin C, Zhang X, Duan X, Xu D, Cheng H, Wang Q, Liu X, Ma J, Ma J (2018) A soil ingestion pilot study for teenage children in China. Chemosphere 202:40–47

Wang Z, Chen J, Chai L, Yang Z, Huang S, Zheng Y (2011) Environmental impact and site-specific human health risks of chromium in the vicinity of a ferro-alloy manufactory, China. J Hazard Mater 190:980–985

Watt J, Thornton I, Cotter-Howells J (1993) Physical evidence suggesting the transfer of soil Pb into young children via hand-to-mouth activity. Appl Geochem 8:269–272

WHO (2000) Air quality guidelines for Europe

WHO (2004) Guidelines for drinking-water quality

WHO (2009) Levels of lead in children's blood

WHO (2011) Guidelines for quality drinking-water

Wongsasuluk P, Chotpantarat S, Siriwong W, Robson M (2014) Heavy metal contamination and human health risk assessment in drinking water from shallow groundwater wells in an agricultural area in Ubon Ratchathani province, Thailand. Environ Geochem Health 36:169–182

Xia Q (2016) Assessing exposure risk of arsenic, cadmium and lead (or mixed with PAHs) in soils using in-vitro methods. Ph.D. Thesis, School of Medicine, The University of Queensland

Xiao J, Wang L, Deng L, Jin Z (2019) Characteristics, sources, water quality and health risk assessment of trace elements in river water and well water in the Chinese Loess Plateau. Sci Total Environ 650:2004–2012

Xiong T, Leveque T, Austruy A, Goix S, Schreck E, Dappe V, Sobanska S, Foucault Y, Dumat C (2014a) Foliar uptake and metal (loid) bioaccessibility in vegetables exposed to particulate matter. Environ Geochem Health 36:897–909

Xiong T, Leveque T, Shahid M, Foucault Y, Mombo S, Dumat C (2014b) Lead and cadmium phytoavailability and human bioaccessibility for vegetables exposed to soil or atmospheric pollution by process ultrafine particles. J Environ Qual 43:1593–1600

Xiong T, Austruy A, Pierart A, Shahid M, Schreck E, Mombo S, Dumat C (2016) Kinetic study of phytotoxicity induced by foliar lead uptake for vegetables exposed to fine particles and implications for sustainable urban agriculture. J Environ Sci 46:16–27

Yang WY, Efremov L, Mujaj B, Zhang ZY, Wei FF, Huang QF, Thijs L, Vanassche T, Nawrot TS, Staessen JA (2018) Association of office and ambulatory blood pressure with blood lead in workers before occupational exposure. J Am Soc Hypertens 12:14–24

Yoon J, Cao X, Zhou Q, Ma LQ (2006) Accumulation of Pb, Cu, and Zn in native plants growing on a contaminated Florida site. Sci Total Environ 368:456–464

Yoshinaga J, Yamasaki K, Yonemura A, Ishibashi Y, Kaido T, Mizuno K, Takagi M, Tanaka A (2014) Lead and other elements in house dust of Japanese residences–Source of lead and health risks due to metal exposure. Environ Pollut 189:223–228

Yu Y, Li Y, Li B, Shen Z, Stenstrom MK (2016) Metal enrichment and lead isotope analysis for source apportionment in the urban dust and rural surface soil. Environ Pollut 216:764–772

Zhang H, Huang B, Dong L, Hu W, Akhtar MS, Qu M (2017) Accumulation, sources and health risks of trace metals in elevated geochemical background soils used for greenhouse vegetable production in southwestern China. Ecotox Environ Safety 137:233–239

Zheng J, Tan M, Shibata Y, Tanaka A, Li Y, Zhang G, Zhang Y, Shan Z (2004) Characteristics of lead isotope ratios and elemental concentrations in PM10 fraction of airborne particulate matter in Shanghai after the phase-out of leaded gasoline. Atmos Environ 38:1191–1200

Zheng J, Chen K, Yan X, Chen SJ, Hu GC, Peng XW, Yuan J, Mai BX, Yang ZY (2013) Heavy metals in food, house dust, and water from an e-waste recycling area in South China and the potential risk to human health. Ecotox Environ Safety 96:205–212

Zheng N, Wang Q, Zheng D (2007) Health risk of Hg, Pb, Cd, Zn, and Cu to the inhabitants around Huludao Zinc Plant in China via consumption of vegetables. Sci Total Environ 383:81–89

Zhuang P, McBride MB, Xia H, Li N, Li Z (2009) Health risk from heavy metals via consumption of food crops in the vicinity of Dabaoshan mine, South China. Sci Total Environ 407:1551–1561

Impact of Lead Contamination on Agroecosystem and Human Health

Vasudev Meena, Mohan Lal Dotaniya, Jayanta Kumar Saha, Hiranmoy Das, and Ashok Kumar Patra

Abstract Environmental threat due to toxic heavy metals is of prime concern worldwide. Rapid urbanization, industrialization and other developmental activities including anthropogenic activities (e.g. mining, fossil fuel burning) are the major contributors of heavy metals above the prescribed permissible limit. High concentration of heavy metals has detrimental impact on soil, water and air as well as human and animal health. Heavy metals exert toxic effects on soil microorganism hence results in the change of the diversity, population size and overall activity of the soil microbial communities. Heavy metal toxicity influences all soil microbial activity that involves change in the microbial population, diversity, their size and growth. Loss of soil fertility results in reduced crop yield and imbalance nutrition due to presence of excess amount of metals. Lead is non-biodegradable highly toxic heavy metal present in the environment. Elevated Pb in soil causes to decrease of soil productivity and impair with various soil enzymatic activities. Lead contaminated soil created several chronic health implications (carcinogenic) or even to death of the living organisms via food chain contamination. Lead is considered as one of the potential carcinogens which can damage cardiovascular, kidney, brain, gastrointestinal tracts, low IQ, loss of hearing or multi-organ failure in humans. Lead toxicity causes several health hazards like everlasting brain injury, hearing loss, learning disabilities, behavioural abnormalities in children while in adults it comes with hypertension, blood pressure, heart disease, and so on. Lead pollution also has severe threat to the aquatic living organisms. In this chapter, a brief overview about the lead with its various active form, source of contamination and impact on agroecosystem, human and animals has been described in details.

V. Meena (✉) · J. K. Saha · H. Das · A. K. Patra
Division of Environmental Soil Science, ICAR-Indian Institute of Soil Science, Bhopal, India

M. L. Dotaniya
ICAR-Directorate of Rapeseed-Mustard Research, Bharatpur, India

© Springer Nature Switzerland AG 2020
D. K. Gupta et al. (eds.), *Lead in Plants and the Environment*, Radionuclides and Heavy Metals in the Environment,
https://doi.org/10.1007/978-3-030-21638-2_4

Keywords Lead · Heavy metals · Agroecosystem · Contaminants · Human health · Toxicity

1   Introduction

Lead (Pb) is one of the most naturally occurring bluish grey widespread highly toxic contaminant in soils which predominantly available in the form of mineral mixed with other elements like sulphur (PbS, $PbSO_4$) or oxygen ($PbCO_3$). The lead concentration in the earth crust ranges from 10 to 30 mg kg^{-1} (USDHHS 1999) while in the soil surface it varies from 10 to 67 mg kg^{-1} with an average value of 32 mg kg^{-1} globally (Kabata-Pendias and Pendias 2001). Among the metals, Pb occupied fifth rank after Fe, Cu, Al and Zn. The soils of municipal areas frequently contain lead concentration more than normal soils because of widespread use of leaded paint and leaded gasoline as well as industrial uses which range from 150 to 10,000 mg kg^{-1} of lead. According to WHO report, estimates mentioned that about 1.43 million cases of death happened due to lead poisoning every year along with 6.0 million new cases of children with intellectual disabilities (WHO 2013). The lead contamination in the soil results in the manifestation of morphological, biological and physiological changes in the plant after its uptake through the roots. Plant adsorbs the lead when the level of soil lead contamination is high. In several countries of North America lead is used in the form of leaded gasoline as a fuel which is the major source of lead pollution. Apart from this, approximately 50% of the total lead is utilized in manufacturing of the storage batteries in Unites States only. Lead is also the key ingredients in many industrial products such as dyes, paints, ceramics, cable covers and bearings (Manahan 2003). In current days, the pollution due to excess of lead has become global potential health concern due to its hazards to human and animals because of its widespread use and environmental contamination. The promising way for the entry of lead to the various systems like plant, animal and human tissues are via air inhalation or ingestion, dietary intake via food chain contamination and manual handling. Airborne contaminants in the environment are released from the motor vehicles. While lead smelter, coal burning and thermal power plants are responsible for aerial pollution in the form soil or dust particles which remains in the atmosphere indefinitely. Furthermore, atmospheric deposition of lead dust particles contaminates the surrounding land and water. The lead-contaminated waste washes off through the runoff during rainy season which contaminates groundwater on leaching of heavy metal toxicants. Different studies reported elevated levels of Pb in blood of children (200 µg l^{-1}) and dogs (250 µg l^{-1}) (Kaul et al. 2003; Balagangatharthilagar et al. 2006). The dumping sites in urban areas are considered as immense contributor to the lead. The highest soil lead level (36.1 mg kg^{-1}) was reported at Aketego dumpsite in Ghana (Twumasi et al. 2016). In India, Patel et al. (2010) reported Pb levels in various environmental media (air, rain water, runoff water, surface soil, sludge and plant) from different industrial sites situated at

Raipur city of Chhattisgarh state of India (cement, steel and ferro-alloy), Bhilai (steel and others) and Korba (thermal power plants and others). These industrial sites reported higher concentration of lead in air and soil. Unfortunately, in urban areas due to shrinkage and pressure on land, many farmers tends to use unapproved land for growing of vegetable crops in dumping sites (Egyir and Beinpuo 2009) which contain plenty of toxic chemicals including heavy toxic metals that may pose risks to the human health when introduces to the human body through the food chain (Bagumire et al. 2009).

2 Source of Lead Contamination

Anthropogenic activities are the primary source of lead contamination which includes industrial and manufacturing developmental processes. In some of the developing countries use of leaded gasoline as fuel also causes atmospheric pollution. Worldwide estimates reveal that about 4.5–5.5 million tons of lead used in gasoline remains in the soil and dust. Municipal sewage and sludge are the major source of multi heavy metals particularly in most developed countries like Europe, America and developing countries like India. The application of sewage in the form of irrigation and sludge to the agricultural land contribute huge amount of the metal like Cu, Zn, Hg, Cd, As, Pb, Co, Ni, Fe, Se and Mn. Phosphorus fertilizers are also another potential source of heavy metal which contains range of metals including lead. Application of phosphate fertilizers in the form of compounds like North Carolina Phosphate Rock (NCPR), Indian phosphate rock, SSP, TSP and DAP bring in lead to the soil. Apart from this, industrial uses of lead such as proofing (also called as proving or blooming, term used by bakers for shaped bread), flashing and sound proofing explore the lead contamination. Further, its multiple daily uses in several industrial products like batteries, paints, pigments, gasoline, ceramics, pesticides and glassware are common (Tangahu et al. 2011). The major sources of lead in soils (Fig. 1) are as follows:

- Batteries
- Metal products
- Industrial waste
- Preservatives
- Petrol additives
- Paints with lead pigments
- Industrial smelting and alloying
- Leaded fuels, bullets and fishing sinkers
- Ceramics
- Dumping sites

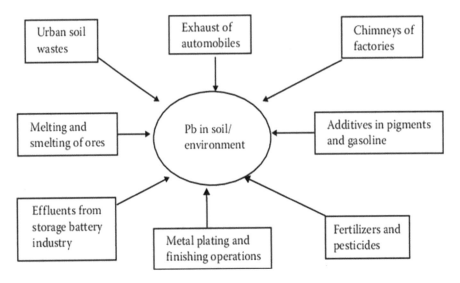

Fig. 1 Schematic representation of various sources of lead contamination to the soil

3 Forms of Lead

Lead oxides and hydroxides are the common form of the lead in the soil. The most stable and reactionary form of lead is Pb (II) and its hydroxyl complexes which form mononuclear and polynuclear oxides and hydroxides (GWRTAC 1997). Primarily the lead (II) compounds exist in its ionic from (e.g. Pb^{2+} $SO_4{}^{2-}$) whereas Pb (IV) compound form covalent like $Pb(C_2H_5)_4$ (Tetraethyl lead) which is the most prominent pollutant in use. Besides this, the other compounds of lead (lead carbonate, lead phosphate, lead hydroxides) are insoluble (Raskin and Ensley 2000). Among the solid forms, lead sulphide is the most common and constant form. In the absence of oxygen microbial alkylation is responsible to convert tetramethyl lead, a volatile organo-lead compound. Lead is also present in salts like $Pb(OH)_2 \cdot 2PbCO_3$ which is commonly used in white paint pigments that cause chronic lead poisoning to children.

4 Behaviour of Lead in Soil

Worldwide lead content in the uncontaminated soil ranges from 10 to 84 ppm with an average value of about 50 ppm. The fate and transport of the lead in the soil after its entry depends upon its chemical form and speciation. After adsorption, it goes through the various mechanisms/processes like leaching through downward movement by surface water during heavy rains, biological mobilization or immobilization, ion exchange, complexation, adsorption and desorption or may get precipitate or taken up by the plants with the help of plant roots. It may also transport to the atmosphere in the form of fine particle through the wind. Sandy soils have low fixing capacity of lead as compare to other soils and it increases with increases of clay and humus content. Soil pH plays crucial role in the existence of lead. In acidic condition (pH < 5) lead is more soluble and held less tightly to the soil, whereas it is the reverse when pH is more (>6.5) with less solubility and stronger binding. About 60–80% or more of dissolved lead present in the form of soil organic complex in uppermost 1–2 in soil surface (Sauve et al. 1997).

5 Impact of Lead Contamination

5.1 Lead Toxicity

Lead toxicity or lead poisoning is also known as plumbism. The level of any metal beyond the permissible limit cause toxic effect due to its poisonous nature and may have pessimistic effect to the life. These toxic metals get accumulated in the body and in the food chain after entry through the various medium. The soil contaminated with lead is categorized into five groups based on the lead contamination level viz. very low (<150 ppm), low (150–400 ppm), medium (400–1000 ppm), high (1000–2000 ppm) and very high (>2000 ppm). The level of lead content varies from season to season and in the different mediums. A study reported that the permissible limit of lead for air (0.10 $\mu g\ m^{-3}$), soil (300 $\mu g\ kg^{-1}$) drinking water (5 $\mu g\ l^{-1}$) and food (1.1 $mg\ kg^{-1}$), respectively. The Pb contamination of the atmospheric air of central India in winter season was reported several folds higher (1.0 $\mu g\ m^{-3}$) than its permissible limit. Likewise, industrial effluents contain higher level of Pb (5 $\mu g\ l^{-1}$) which is highly toxic to human, animals as well as for agricultural purposes, if used as irrigation water. The crops cultivated with the sewage effluents contain much higher level of Pb (1.1 $mg\ kg^{-1}$) in the vegetative parts (Patel et al. 2010).

5.2 Soil Health and Soil Biodiversity

Soil is an important entity which is made up of three different components viz. lithosphere, hydrosphere and atmosphere and disturbances of any one of among these severely affects the soil health with its phenomenon. As discussed earlier in the text that soil is the most vulnerable to the contamination by various sources including heavy metals which harshly harm the soil health that results in the loss of crop yield by about 15–25%. Soil is the major source of contamination of food chain due to presence of heavy metals like Cr, Co, Hg, Pb, As and Zn. Fertilizers viz. rock phosphate or phosphatic fertilizers responsible for soil and food crops contamination in the form of impurities (Gupta et al. 2014). Long term fertilizer experiments (>39 years) confirmed build-up of heavy metal (Cd, Pb, Co, Cr and Ni) in the soil due to 100% or more application of recommended dose of NPK (Adhikari et al. 2012). Similarly, long term sewage irrigation results in significant build-up of lead in upland soil at Kolkata (Adhikari et al. 1993). Higher concentration of lead (8–16 times above the reference limit) was found in the crops grown on polluted area near the lead smelter plant in France (Douay et al. 2013). Use of sewage water for growing of crops had drastic effect on beneficial soil micro-organisms, hammering soil fertility, imbalance nutrition and productivity which modify the cropping pattern of the farming.

The agricultural soils/lands which situated nearby to the industrial areas are more prone to the heavy metal contamination (Saha 2013). The lead concentration was higher in the soil near to electroplating and paint industry at Coimbatore which considered as the major source of lead contamination. The cement factory at Dindigul district of Tamil Nadu is the major contributor of lead dust pollution up to 1.0 km of distance that was more prominent within the radius of half km (Stalin et al. 2010). The surrounding areas of thermal power plant were enriched with high accumulation of heavy metals like As, Hg, Ni and Cr (Ozkul 2016) and the Soil and plant samples were detected with high geo-accumulation index and enrichment factor. Atmospheric deposition of heavy metals nearby industrial sites and heavy traffic areas are also reported to high lead accumulation. A lead smelter plant emitted about 18.4 tons of Pb, 26.2 tons Zn, 823 kg Cd in 2001 in France which results in high concentration of Cd, Pb and Zn accumulation in top 30 cm surrounding soil (Godin et al. 1985; Sterckeman et al. 2000, 2002). Even after closure of the plant in 2003, the soil particles of lawn and kitchen gardens had about 47–68% of human bioaccessible heavy metals including lead on soil or dust particle ingestion (Pelfrene et al. 2013). Furthermore, the soils of the nearby areas of the smelter plants had 10–15 times more lead content than the reference soils (Pruvot et al. 2006).

The high accumulation of the lead in the soil is detrimental to biological life of the soil. Lead toxicity reduces the population of the beneficial micro-organisms like earthworm, bacteria, fungi, protozoa and arthropods. which are responsible for improved soil fertility and other soil biological function. Due to inhibition of growth and morphology of beneficial micro-organisms, different microbial assisted soil processes of agricultural importance get hampered like biological nitrogen fixation by influencing *Rhizobia* (Hernandez et al. 2003). Soil microbial diversity is the key component for the maintenance of the soil quality (Xie et al. 2014). The soil lead accumulation affected both soil enzymatic activities as well as soil microbial diversity.

5.3 Plant Growth and Development

Soil friability and soil compaction are the two important soil quality parameters that influence overall plant growth and development which ultimately linked to final output, that is, crop yield. Pollutants can affect the plant by both ways as directly as

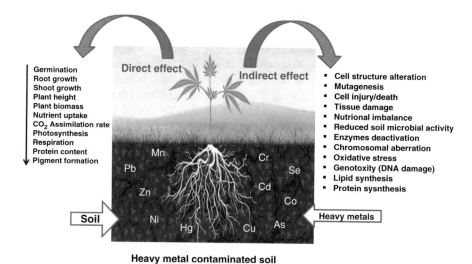

Fig. 2 Soil lead contamination in plant and their effect on growth and development

well as indirectly due to contaminated soil and water use (Fargasova 1994). Plant absorbs the metals and other pollutants from the soil with the help of their root system in ionic form (Fig. 2). Under high concentration of metals and other pollutants plant adsorb the toxic elements which hamper the seed germination, plant growth and development by hindering the physiological, biochemical and genetic elements of the plant (Sethy and Ghosh 2013). Further toxicants disturb the proper functioning of cell and cell constituents, cell wall, proteins and polynucleotides. Plant exhibits toxicity only when soil lead level is high. Studies reported that lead get accumulated more in the leafy vegetables (lettuce, spinach, etc.), whereas a little amount of lead may accumulate in fruiting parts and fruits.

Studies reported the safe limit of soil lead level (up to 300 ppm) and the risk of lead poisoning is increases with increasing level of lead content (Rosen 2002) due to soil or dust deposition rather than plant uptake. Organic matter and phosphorus content in the soil antagonize the soil lead content and its risk to the plant. Under high soil lead concentration, the plants enzymes get deactivate and also disrupt the calcium metabolism which reduces CO_2 assimilation and finally photosynthesis gets affected.

5.4   Crop Quality and Productivity

Cultivation of the crops in soils contaminated with the heavy metals results in the loss of crop quality with reduced productivity. The bioaccumulation of lead in the crops (Grain and straw/biomass) reduced quality of the food grain and has toxic effect on health of human and animals when consumed. The vegetables and their parts accumulated more toxic metals than the edible grain crops. The vegetable crops grown with the continuous use of sewage water nearby urban or peri-urban areas had high level of the lead content (Bhupal Raj et al. 2009) which reduced the shelf life of the crops with giving bad odour. The inferior quality of onion was found with polluted groundwater in Ratlam (India) industrial area (Saha and Sharma 2006); vegetable samples collected from Yamuna flood plains Near Delhi, Faridabad and surrounding areas had high content of Pb and Zn (DFID 2003). Among the crops, spinach is the most sensitive crop for the heavy metal accumulation including lead. About 72% of spinach samples collected from Yamuna flood plains had lead concentration above the permissible limit (2.5 mg kg^{-1}) and 24% sample contain lead >5.0 mg kg^{-1}. The groundwater samples collected from polluted area of Nagda (India) contain 9.1 µg l^{-1} of Pb which was 162-folds higher than the unpolluted groundwater samples. Studies reported bioaccumulation of heavy metals in potato, rice, wheat at local and international commodity marketing and more concern on horticultural product which contained high level of Cu due to heavy use of copper containing fungicide. Similarly, accumulation of Pb resulting from the use of leaded fuels is widespread worldwide including India (Bolan et al. 2003).

Due to scarcity of fresh water for irrigation, the farmers nearby urban and peri-urban areas are forced to use the sewage water or industrial effluents. This wastewater contains enormous amounts of toxic elements including heavy metals which directly or indirectly discharged into water bodies and thus pollution of groundwater. Use of groundwater for irrigation adversely affected the farmer's economic condition besides changing cropping patter and reduced crop productivity (Saha and Sharma 2006). The farmers moved out from growing of pulses and vegetables with the polluted groundwater and kept their land fallow in the rainy season. The severe losses in the crop yield of soybean, gram, fenugreek and garlic have been observed. The crops also get infected with more pest and diseases due to contaminated water which increased input cost to the farmers. The groundwater sample collected from Pithampur (MP) industrial area contain 3.7 µg Pb l^{-1} and effluents samples from Patancheru (AP) contain 0.3–14.2 µg Pb l^{-1} (Panwar et al. 2010).

5.5 Nutrients Availability and Enzymatic Activities in Soil

Degradation of soil health induced due to the soil pollutant hamper the nutrient availability in the soil and reduces the fertility of the soil by imbalance nutrition. The heavy metals or metalloids and other toxic elements antagonize the availability of beneficial nutrients to the crops. For example, elevated Pb lowers the availability of the calcium and phosphorus by which calcium and phosphorus mechanism affected. Lead had strong inverse correlation with iron status. Deficiency of zinc also induced more absorption of the lead (Markowitz and Rosen 1981). Highly lead contaminated soils found to be more deficient in iron, calcium, phosphorus and zinc. Nutrient cycling is the biochemical process mediated by a variety of the soil micro-organisms and all the processes or reactions are catalysed by various enzyme groups (Tabatabai 1982). The excess of heavy metals obstruct various soil enzymatic activities (Yang et al. 2006) like alkaline phosphatases, dehydrogenase, β-glucosidase, protease, urease and cellulose (Kunito et al. 2001; Effron et al. 2004; Oliveira and Pampulha 2006; Wang et al. 2008). A study done by Saha (2013) reported that enzymatic activities have declined with increasing level of metals.

5.6 Human and Animal Health

In recent days the heavy metal contamination of various mediums is of great concern globally. The metal contamination not only harms the soil, water and atmospheric environment but also has adverse impact on the human as well as animal health (Singh and Ghosh 2011). Heavy metal toxicity to the human health received attention due to their widespread poisoning which results in "Gasio gas" (trimethyl arsine) due to arsenic toxicity and secondly "Minamata disease" in Japan in late 1950s due to mercury (Hg) through ingestion of fish. Millions of people were also affected due to heavy metal poisoning or toxicity in China, Bangladesh including India. In Australia and New Zealand accumulation of cadmium in the body of grazing animals were reported due to that meat product were totally banned to overseas markets.

Generally, the heavy metal in the human body get enters through the two ways, that is, inhalation and ingestion. The transfer of heavy metal through the food chain start from the contaminated soil by plant root uptake and further transfer to the edible parts like grain or fruits which end to human or animal body. The entry of the metals to the food chain depends on source and amount of metal, plant type, climatic condition, seasonal variation, soil pH, and so on. Long-term use of sewage sludge contaminated soil for crop production results in accumulation of high concentrations of Cu, Co, Cd, Hg, Pb, As and Zn, in many countries and their subsequent adverse impact on human and animal health (ATSDR 2005). Heavy metals like Co, Cr, Ni and Cu have a more toxic effect on plants than animals, whereas Pb, Cd, Hg and As are more poisonous to higher animals (McBride 1994). The lead in the body enters through the breathing or swallowing of soil or dust particle and cause detrimental health implications by its accumulation in the body tissues. Children are more susceptible to lead toxicity than adults. Younger persons and fasting individuals ingested more amount of lead. The absorption rate of lead in the bloodstream ranges from 3% to 80% (FAO/WHO 2011; Sabath and Robles-Osorio 2012; Marsh and Bailey 2013) and half-life of Pb in blood and tissues considered about 28–36 days (Farzin et al. 2008).

The adsorption of lead in children occurs mainly through the inhalation of the dust particles or via ingestion of contaminated food and water. Insufficiency of Ca, Fe and P or imbalance nutrition in the body and empty stomach increases the uptake of lead. Lead accumulates in the body organs like brain, bones and inhibits their proper functioning by hampering haem and haemoglobin formation may lead to plumbism or even death. Concentration of lead in body above the permissible limit may cause serious injury to RBC, kidney, brain, nervous system and gastrointestinal tract (Baldwin and Marshall 1999), effects on renal, cardiovascular, reproductive system (Patra et al. 2011; FAO/WHO 2011; Abdullahi 2013; Sun et al. 2014), further loss of IQ, hearing, mental deterioration, hypersensitivity, impaired neurobehavioral functioning (Grandjean and Landrigan 2014) and many developmental disorder may come across in children below 6 years of age. Increasing lead concentration also impaired with the headache, vitamin D metabolism, loss of appetite, constipation, poor haemoglobin synthesis, frank anaemia, nephrotoxic effects with impaired renal excretion of uric acid, nephrotoxic effects with impaired renal excretion of uric acid and weakness of joints (NSC 2009).

Studies reported that provisional tolerable weekly intake (PTWI) of 25000 mg kg^{-1} body weight of lead reduces the IQ by three points in children and increases systolic BP by 3 mmHg in adults (0.4 kPa) in adults (WHO/FAO 2011; Tellez-Plaza et al. 2012; Solenkova et al. 2014). The epidemiological studies results from lead smelter company of northern France clear that high blood lead levels (39.5 µg l^{-1}) of children living on the polluted sites as compare to the unpolluted sites (30.6 µg l^{-1}) (Leroyer et al. 2000, 2001), among them 10–15% had higher blood lead level of 100 µg Pb l^{-1} (Sterckeman et al. 2000, 2002; Pelfrene et al. 2013). Similarly, rice grain contained about 0.39 mg kg^{-1} of lead beyond maximum permissible limit (0.2 mg kg^{-1}) due to fly ash use in China (Nolan 2003; Türkdogan et al. 2003). Being as constitute of azo-dyes for colouring may results in carcinogenic effect on human body.

5.7 Aquatic Organisms

Heavy metal pollution causes major effect on biotic life cycle (Meena et al. 2014). Increasing the concentration of toxic substances in the tissue of the tolerant organisms at higher level called as biomagnifications or bioamplification. Due to movement of pollutants (heavy metals) into water bodies from different diffuse or point sources has adverse impact on aquatic organisms. Bioaccumulation of heavy metals in the aquatic system posing a severe menace to the aquatic fauna and flora particularly to fishes which constitutes one of the major sources of protein rich food for mankind. The industrial waste in the form of sewage and sludge (e.g. effluents, wastewater) and municipal waste discharged directly with or without any treatment that constitutes number of toxic elements including heavy metals which pollute the water bodies (ponds, lakes, rivers, etc.) and groundwater as well. The polluted water impedes the growth and development of water living organisms that may even lead to the death to the aquatic life. These toxic heavy metals may reach human bodies via the food chain (e.g. fish, prawns, crabs that cause severe health hazards, like "Minamata disease" in Japan in late 1950s due to transfer of methyl mercuric compound). Due to industrialization development, the pollution of aquatic environment has been increased by two- to threefold more as compared to pre-industrial levels.

5.8   Environmental Threats

Heavy metals are carcinogenic in nature; therefore, its decontamination from the system is necessary for sustainable crop production or a healthy environment (Dotaniya et al. 2018). Huge volume of the wastewater discharged from the textile industries using wide variety of chemicals causing severe damage to the environment. In India, thermal power plants emit 339 t of Pb annually along with other heavy metals (Pacyna and Pacyna 2001). Atmospheric release of Pb from the thermal plants causes air pollution and pollution of water streams and groundwater due to discharge of their effluents. Lead is such toxic heavy metal which can stay behind in the environment permanently in the form of fine dust particle. Use of gasoline as a fuel is the major contributor of lead to the air pollution. Another by-product of the thermal plant is fly ash, a hazardous material which contains a variety of heavy metals and is a threat to the whole environment. Fly ash contains a variety of heavy metals which contaminate the aerial environment along with surrounding agricultural soil and water bodies; upon settling after long exposure, these toxic metals enter the food chain. Unplanned urbanization, infrastructure development and industrialization have resulted in atmospheric pollution due to deposition of heavy metals as dust particle from various anthropogenic activities like mining and other urban developmental activities (Mishra et al. 2013; Pal et al. 2014). In Varanasi (India), the atmospheric deposition of lead was reported from 9.8 g ha^{-1} year^{-1} Sharma et al. (2008) to as high as 71.0 g ha^{-1} year^{-1} (Tiwari et al. 2008) and similarly, that was 32.4 g ha^{-1} year^{-1} for Mumbai (Tripathi et al. 1993). The emitted particulate matter from smelting industries contains oxides of heavy metal such as Zn, Cd, Pb, AS, Hg, Cu which are harmful to the whole surrounding environment. The smelting processes produce about 3 tons of solid waste for the production of one ton of lead or zinc which contain about 0.5–0.7% lead along with other toxic heavy metals.

5.9   Impact on Agroecosystem

Anthropogenic activities (e.g. mining) play crucial role in the transfer of the toxicants to the soil, water and air from the underground bed rocks. Mining activities introduces variety of heavy metals (Cr, Cu, Cd, Pb, Hg, As, Co and Zn) which down streamed through the leaching and runoff in rainy season after mineralization process and contaminate the groundwater resources. The heavy metals are released into the various environment on weathering of these bed rocks and minerals which ultimately contaminant the surrounding ecosystem. These toxic metal(oid)s moved out to the agricultural land/soil together with runoff water or after using sewage water as irrigation. The soil and vegetable sample collected from mining areas were highly contaminated with Pb, Cd and As while irrigation water contain high level of Pb.

More than 71% of the vegetable samples were found heavily contaminated with lead in the mining areas. Similarly 44% samples were with high As content. Heavy metals released from the smelter industries get deposited in the terrestrial ecosystem which causes agricultural land and food chain contamination. Similar results also reported from North Vietnam due to extensive mining activities to deteriorate soil, water bodies and plants with heavy metals (Bui et al. 2016).

6   Summary and Conclusion

Lead is an extremely toxic and hazardous heavy metal contaminant even at low concentrations to living organisms. Due to its non-biodegradable nature it can be accumulating in the living organisms, thus causing several health risks. Proper disposal of the lead-based waste products should be strictly followed to avoid the environmental contamination or further exposure. Remediation of lead contamination is indispensable to reduce the associated potential risks to various environments and to make land resources lead free for agricultural production, to make available healthy food for mankind. A summary of the chapter is represented with the help of a schematic representation (Fig. 3).

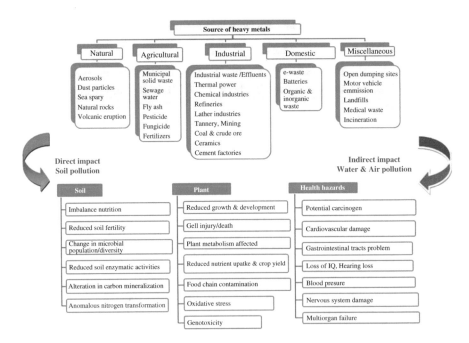

Fig. 3 Schematic representation of lead contamination and its fate in the soil environment

References

Abdullahi M (2013) Toxic effects of lead in humans: an overview. Glob Adv J Environ Sci Toxicol 2:157–162

Adhikari S, Gupta SK, Banerjee SK (1993) Heavy metals content of city sewage and sludge. J Indian Soc Soil Sci 41:170–172

Adhikari T, Wanjari RH, Biswas AK (2012) Final Report of the project entitled "Impact assessment of continuous fertilization on heavy metals and microbial diversity in soils under long-term fertilizer experiment". Submitted to Ministry of Forest and Environment, New Delhi, India. p 175

ATSDR (2005) Toxicological profile for nickel. Agency for Toxic Substances and Disease Registry, U.S. Department of Health and Human Services, Public Health Service, Division of Toxicology/Toxicology Information Branch, Atlanta, Georgia

Bagumire A, Todd EC, Nasinyama GW, Muyanja C, Rumbeiha WK, Harris C, Bourquin LD (2009) Potential sources of food hazards in emerging commercial aquaculture industry in sub-Saharan Africa: a case study for Uganda. Int J Food Sci Tchnol 44:1677–1687

Balagangatharthilagar M, Swarup D, Patra RC, Diwedi SK (2006) Blood lead level in dogs from urban and rural areas of India and its relation to animal and environmental variables. Sci Total Environ 359:130–134

Baldwin DR, Marshall WJ (1999) Heavy metal poisoning and its laboratory investigation. Anna Clin Biochem 36:267–300

Bhupal Raj G, Singh MV, Patnaik MC, Khadke KM (2009) Four decades of research on micro and secondary-nutrients and pollutant elements in Andhra Pradesh. Research Bulletin, AICRP Micro- and Secondary-Nutrients and Pollutant Elements in Soils and Plants, IISS, Bhopal, India. p. 1–132

Bolan NS, Adriano DC, Mani S, Khan AR (2003) Adsorption, complexation and phytoavailability of copper as influenced by organic manure. Environ Toxicol Chem 22:450–456

Bui ATK, Nguyen HTH, Nguyen MN, Tran TH, Vu TV, Nguyen CH, Reynolds HL (2016) Accumulation and potential health risks of cadmium, lead and arsenic in vegetables grown near mining sites in Northern Vietnam. Environ Monit Assess 188:525

DFID (2003) Heavy metal contamination of vegetables in Delhi. Executive summary of technical report. Department for International Development, London, U.K.

Dotaniya ML, Panwar NR, Meena VD, Dotaniya CK, Regar KL, Lata M, Saha JK (2018) Bioremediation of metal contaminated soil for sustainable crop production. In: Meena VS (ed) Role of rhizospheric microbes in soil. Springer, Singapore, p. 143–173

Douay F, Pelfrene A, Planque J, Fourrier H, Richard A, Roussel H, Girondelot B (2013) Assessment of potential health risk for inhabitants living near a former lead smelter. Part 1: metal concentrations in soils, agricultural crops, and homegrown vegetables. Environ Monit Assess 185:3665–3680

Effron D, de la Horra AM, Defrieri RL, Fontanive V, Palma PM (2004) Effect of cadmium, copper, and lead on different enzyme activities in a native forest soil. Comm Soil Sci Plant Anal 35:1309–1321

Egyir I, Beinpuo E (2009) Strategic innovations in urban agriculture, food supply and livelihood support systems performance in Accea, Ghana. College of Agriculture and Consumer Sciences. University of Ghana.

FAO/WHO (2011) Joint FAO/WHO Food Standards Programme Codex Committee On Contaminants In Foods. Available at: ftp://ftp.fao.org/codex/meetings/CCCF/cccf5/cf05_INF. pdf

Fargasova A (1994) Effect of Pb, Cd, Hg, As, and Cr on germination and root growth of *Sinapis alba* seeds. Bull Environ Cont Toxicol 52:452–456

Farzin L, Amiri M, Shams H, Faghih MAA, Moassesi ME (2008) Blood levels of lead, cadmium, and mercury in residents of Tehran. Biol Trace Elem Res 123:14–26

Godin P, Feinberg M, Ducauze C (1985) Modelling of soil contamination by airborne lead and cadmium around several emission sources. Environ Pollut 10:97–114

Grandjean P, Landrigan PJ (2014) Neurobehavioural effects of developmental toxicity. Lancet Neurol 13:330–338

Gupta DK, Chatterjee S, Datta S, Veer V, Walther C (2014) Role of phosphate fertilizers in heavy metal uptake and detoxification of toxic metals. Chemosphere 108:134–144

GWRTAC (1997) Remediation of metals-contaminated soils and groundwater, Tech. Rep. TE-97-01, GWRTAC, Pittsburgh, Pa, USA, GWRTAC-E Series

Hernandez L, Probst A, Probst JL, Ulrich E (2003) Heavy metal distribution in some French forest soils: evidence for atmospheric contamination. Sci Total Environ 312:195–219

Kabata-Pendias A, Pendias H (2001) Trace metals in soils and plants. CRC Press, Boca Raton, FL

Kaul PP, Shyam S, Srivasatva R, Misra D, Salve PR, Srivastava SP (2003) Lead levels in ambient air and blood of pregnant mothers from the general population of Lucknow (UP), India. Bull Environ Cont Toxicol 71:1239–1243

Kunito T, Saeki K, Nagaoka K, Oyaizu H, Matsumoto S (2001) Characterization of copper resistant bacterial community in rhizosphere of highly copper-contaminated soil. Eur J Soil Biol 37:95–102

Leroyer A, Nisse C, Hemon D, Gruchociak A, Salomez JL, Haguenoer JM (2000) Environmental lead exposure in a population of children in northern France: factors affecting lead burden. Amer J Ind Med 38:281–289

Leroyer A, Hemon D, Nisse C, Auque G, Mazzuca M, Haguenoer JM (2001) Determinants of cadmium burden levels in a population of children living in the vicinity of non-ferrous smelters. Environ Res 87:147–159

Manahan SE (2003) Toxicological chemistry and biochemistry. CRC Press, Boca Raton, FL

Markowitz ME, Rosen JF (1981) Zinc and copper metabolism in CaNa$_2$ EDTA-treated children with plumbism. Pediat Res 15:635

Marsh J, Bailey M (2013) A review of lung–to–blood absorption rates for radon progeny. Radiat Prot Dosimetry 157:499–514

McBride MB (1994) Environmental chemistry of soils. Oxford University Press, New York

Meena VD, Dotaniya ML, Coumar V, Rajendiran S, Kundu S, Rao AS (2014) A case for silicon fertilization to improve crop yields in tropical soils. Proc Natl Acad Sci India, Sect B Biol Sci 84:505–518

Mishra AK, Maiti SK, Pal AK (2013) Status of PM10 in bound heavy metals in ambient air in certain parts of Jharia coal field, Jharkhand, India. Int J Environ Sci 4:141–150

Nolan K (2003) Copper toxicity syndrome. J Orthomol Psych 12:270–282

NSC (2009) Lead Poisoning, National Safety Council, 2009, http://www.nsc.org/newsresources/Resources/Documents/Lead Poisoning.pdf

Oliveira A, Pampulha ME (2006) Effects of long-term heavy metal contamination on soil microbial characteristics. J Biosci Bioeng 102:157–161

Ozkul C (2016) Heavy metal contamination in soils around the Tunçbilek thermal power plant. Environ Monit Assess 188:284

Pacyna JM, Pacyna EG (2001) An assessment of global and regional emissions of trace metals to the atmosphere from anthropogenic sources worldwide. Environ Rev 9:269–298

Pal R, Mahima A, Tripathi A (2014) Assessment of heavy metals in suspended particulate matter in Moradabad. India. J Environ Biol 35:357–361

Panwar NR, Saha JK, Adhikari T (2010) Soil and water pollution in India: some case studies. IISS Technical Bulletin. Indian Institute of Soil Science, Bhopal, India

Patel KS, Ambade B, Sharma S, Sahu D, Jaiswal NK, Gupta S, Dewangan RK, Nava S, Lucarelli F, Blazhev B, Stefanova R, Hoinkis J (2010) Lead environmental pollution in Central India. In: Ramov B (ed) New trends in technologies. InTech. Available from: http://www.intechopen.com/books/new-trends-in-technologies/lead-environmental-pollution-in-central-India

Patra R, Rautray AK, Swarup D (2011) Oxidative stress in lead and cadmium toxicity and its amelioration. Vet Med Int:457327

Pelfrene A, Douay F, Richard A, Roussel H, Girondelot B (2013) Assessment of potential health risk for inhabitants living near a former lead smelter. Part 2: Site-specific human health risk assessment of Cd and Pb contamination in kitchen gardens. Environ Monit Assess 185:2999–3012

Pruvot C, Douay F, Herve F, Waterlot C (2006) Heavy metals in soil, crops and grass as a source of human exposure in the former mining areas. J Soils Sedim 6:215–220

Raskin I, Ensley BD (2000) Phytoremediation of toxic metals: Using plants to clean up the environment. John Wiley & Sons, New York, U.S.A.

Rosen CJ (2002) Lead in the home garden and urban soil environment, Communication and Educational Technology Services, University of Minnesota Extension. https://conservancy.umn.edu/bitstream/handle/11299/93998/1/2543.pdf

Sabath E, Robles-Osorio ML (2012) Renal health and the environment: heavy metal nephrotoxicity Medio ambiente y riñón: nefrotoxicidad por metales pesados. Nefrologia 32:279–286

Saha JK (2013) Risk assessment of heavy metals in soil of a susceptible agroecological system amended with municipal solid waste compost. J Indian Soc Soil Sci 61:15–22

Saha JK, Sharma AK (2006) Impact of the use of polluted irrigation water on soil quality and crop productivity near Ratlam and Nagda industrial area. IISS Bulletin, Indian Institute of Soil Science, Bhopal, India. p. 26

Sauve S, McBride M, Hendershot W (1997) Speciation of lead in contaminated soils. Environ Pollut 98:149–155

Sethy SK, Ghosh S (2013) Effect of heavy metals on germination of seeds. J Nat Sci Biol Med 4:272–275

Sharma RK, Agrawal M, Marshall FM (2008) Atmospheric deposition of heavy metal (copper, zinc, cadmium and lead) in Varanasi city, India. Environ Monit Assess 142:269–278

Singh SK, Ghosh AK (2011) Entry of arsenic into food material—a case study. World Appl Sci J 13:385–390

Solenkova NV, Newman JD, Berger JS, Thurston G, Hochman JS, Lamas GA (2014) Metal pollutants and cardiovascular disease: mechanisms and consequences of exposure. Amer Heart J 168:812–822

Stalin P, Singh MV, Muthumanickam D (2010) Four decades of research on micro and secondary nutrients and pollutant elements in crops and soils of Tamil Nadu. Research Publication No. 8. AICRP Micro- and Secondary-Nutrients and Pollutant Elements in Soils and Plants, IISS, Bhopal, India

Sterckeman T, Douay F, Proix N, Fourrier H (2000) Vertical distribution of Cd, Pb and Zn in soils near smelters in the north of France. Environ Pollut 107:377–389

Sterckeman T, Douay F, Proix N, Fourrier H, Perdrix E (2002) Assessment of the contamination of cultivated soils by eighteen trace elements around smelters in the north of France. Water Air Soil Pollut 135:173–194

Sun HJ, Rathinasabapathi B, Wu B, Luo J, Pu LP, Ma LQ (2014) Arsenic and selenium toxicity and their interactive effects in humans. Environ Int 69:148–158

Tabatabai MA (1982) Soil enzymes. In: Page AL, Miller RH, Keeney DR (eds) Methods of soil analysis, Part 2, Chemical and microbiological properties, Monograph no, vol 9. ASA-SSSA, Madison, pp 903–947

Tangahu BV, Sheikh Abdullah SR, Basri H, Idris M, Anuar N, Mukhlisin M (2011) A review on heavy metals (As, Pb, and Hg) uptake by plants through phytoremediation. Int J Chem Eng 2011:1–32

Tellez-Plaza M, Navas-Acien A, Menke A, Crainiceanu CM, Pastor-Barriuso R, Guallar E (2012) Cadmium exposure and all-cause and cardiovascular mortality in the US general population. Environ Health Persp 120:1017–1022

Tiwari K, Pandey A, Pandey J (2008) Atmospheric deposition of heavy metals in a seasonally dry tropical urban environment (India). J Environ Res Develop 2:605–611

Tripathi RM, Ashawa SC, Khandekar RN (1993) Atmospheric depositions of Cd, Pb, Cu and Zn in Bombay, India. Atmos Environ 27:269–273

Türkdogan MK, Fevzi K, Kazim K, Ilyas T, Ismail U (2003) Heavy metals in soil, vegetables and fruits in the endemic upper gastrointestinal cancer region of Turkey. Environ Toxicol Pharm 13:175–179

Twumasi P, Tandoh MA, Borbi MA, Ajoke AR, Tenkorang EO, Okoro R, Dumevi RM (2016) Assessment of the levels of cadmium and lead in soil and vegetable samples from selected dumpsites in the Kumasi Metropolis of Ghana. Afr J Agric Res 11:1608–1616

USDHHS (1999) Toxicological profile for lead, United States Department of Health and Human Services, Atlanta, GA.

Wang Y, Li Q, Shi J, Lin Q, Chen XC, Wu W, Chen YX (2008) Assessment of microbial activity and bacterial community composition in the rhizosphere of a copper accumulator and a non-accumulator. Soil Biol Biochem 40:1167–1177

WHO (2013) Health topics. Available at http://www.emro.who.int/health-topics-section/

Xie Y, Luo H, Du Z, Hu L, Fu J (2014) Identification of cadmium- resistant fungi related to Cd transportation in bermuda grass [*Cynodon dactylon* (L.) Pers.]. Chemosphere 117:786–792

Yang Z, Liu S, Zheng D, Feng S (2006) Effects of cadmium, zinc, and lead on soil enzyme activities. J Environ Sci 18:1135–1141

Lead Contamination and Its Dynamics in Soil–Plant System

M. L. Dotaniya, C. K. Dotaniya, Praveen Solanki, V. D. Meena, and R. K. Doutaniya

Abstract Heavy metal pollution is emerging at a faster rate in non-contaminated areas, which are near to metropolitan cities. The ever-growing population with higher growth rate, industrial extension and poor management of natural resources across the globe in developing countries, particularly in India, make environments vulnerable to metal contamination. Over the last few decades, owing to scarcity of good quality water, farmers are forced to utilize marginal quality water for irrigation purpose which has led to reduction in soil health crop production potential. Long-term application of industrial effluent are accumulating significant amounts of heavy metals in soils and reach to human body via food chain contamination. It causes different types of ill effect in human being. Among heavy metals, lead (Pb) occupies the top place as priority pollutant as per classification given by the Central Pollution Control Board. Due to its toxicity in human body, poor development of infants, mental weakness, nerve disorder, kidney damage, gastric problem, hormonal imbalance and cardiovascular disease are common. The main sources of Pb contamination are acid batteries, paint and varnish, coal industries, automobile sectors, plastic and agriculture sectors. Lack of proper technological intervention in respect to waste water treatment or its safe disposal, enhances the more chance of Pb contamination in soil and water bodies. In nature, it is present as Pb^{2+}, different

M. L. Dotaniya (✉)
ICAR-Directarate of Rapeseed-Mustard Research, Bharatpur, Rajasthan, India
e-mail: Mohan.Dotaniya@icar.gov.in

C. K. Dotaniya
College of Agriculture, SKRAU, Bikaner, Rajasthan, India

P. Solanki
Department of Environmental Science, GBPUA&T, Pantnagar, Uttarakhand, India

V. D. Meena
ICAR-Indian Institute of Soil Science, Bhopal, Madhya Pradesh, India

R. K. Doutaniya
OPJS University, Churu, Rajasthan, India

© Springer Nature Switzerland AG 2020
D. K. Gupta et al. (eds.), *Lead in Plants and the Environment*, Radionuclides and Heavy Metals in the Environment,
https://doi.org/10.1007/978-3-030-21638-2_5

types of oxide and hydroxides, and oxyanionic complexes; their chemistry is affected by various soil and water parameters. Safe disposal and remediation of Pb-contaminated soils by using traditional and modern tools and techniques is the need of the day for sustainable crop production.

Keywords Climate change · Heavy metals · Lead toxicity · Lead–carbon dynamics · Phytoremediation · Soil microbial diversity

1 Introduction

Growing crop plants on healthy soil produces quality foodstuffs. Population increasing at a staggering rate needs more amount of food to alleviate hunger. India is a developing country with second place in population pressure after China. Using poor-quality water or contaminated lands is paramount for producing surplus food owing to lack of healthy soil and fresh water (Bharti et al. 2017). The natural resources are shrinking with time and shrinking fresh water availability is tinted everywhere across the globe. Growing industrial sector and related anthropogenic activities are producing a lot of wastes in terms of solid and effluents (Aktar et al. 2009; Rajendiran et al. 2015; Solanki et al. 2018), which are having huge water potential and meagre amount of plant nutrients. In developing countries most of household waste merging in industrial waste, and reach to farmers fields (Meena et al. 2015; Meena et al. 2019a). In the present time, it is gaining a lot of importance at national and international level for sustainable management of heavy metals, including lead (Pb) toxicity without compromising ecosystem services. Researchers are tirelessly working on minimizing the Pb contamination in food chain with the help of organic and inorganic sources (Bell and MacLeod 1983). The availability of Pb metal in soil for plant roots is reduced by adding organic matter (OM) or pH-mediating substances (Saha et al. 2017).

On the other hand, soil contains a range of heavy metals, which are carcinogenic, and adversely affects the soil–plant nutrient dynamics and soil health (Rajendiran et al. 2018). These metal(s) or metalloid(s) have anatomic density greater than $5 \, \text{gcc}^{-1}$ or atomic number greater than calcium (Dotaniya et al. 2018c). Most common heavy metals are chromium (Cr), lead (Pb), arsenic (As), selenium (Se), mercury (Hg), cadmium (Cd), iron (Fe) and nickel (Ni). Among heavy metals, Pb has occupies a prominent place and is mainly contributed by paint, petroleum combustion, battery-based industries (Table 1). As per international research organizations, lead is categorized as a potential pollutant and found to easily accumulate in soil and sediments containing significant amounts of clay or organic matter. Deposition in soil is regulated by soil properties like pH, clay percent, cation exchange capacity, organic matter content and soil structure. Higher amounts of clay accumulate more amounts of Pb and the soil health declines. Most of the heavy metals are more available at lower pH conditions and precipitate with other minerals at a higher pH

Table 1 Source of heavy metals in environment (Adopted from Levinson (1974) and Alloway (1990)

Metalliferous mining and smelting
Spoil heaps and tailing-contamination through weathering, wind erosion (As, Cd, Hg, Pb)
Fluvially dispersed tailings-deposited on soil during flooding, river dredging, etc. (As, Cd, Hg, Pb)
Transported ore separates-blow from conveyance onto soil (As, Cd, Hg, Pb)
Smelting-contamination due to wind erosion in industrial belt (As, Cd, Hg, Pb, Sb, Se, Fe, Mn)
Iron and steel industries (Cu, Ni, Pb)
Metal finishing (Zn, Cu, Ni, Cr, Cd)
Industry
Plastics (Co, Cr, Cd, Hg)
Textiles (Zn, Al, Z, Ti, Sn)
Microelectronics (Cu, Ni, Cd, Zn, Sb)
Wood preserving (Cu, Cr, As)
Refineries (Pb, Ni, Cr)
Atmospheric deposition
Industrial deposition (Cd, Cu, Fe, Pb, Sn, Hg, V)
Pyrometallurgical industries (As, Cd, Cr, Cu, Mn, Ni, Pb, sb, Ti, Zn)
Automobiles exhausts (Mo, Pb, V)
Fossil fuel combustion (As, Pb, Sb, Se, U, V, Zn, Cd)
Agriculture
Fertilizers (in P fertilizers-Cd, As, V)
Manures (Pig and poultry-As, Cu; FYM-Mn and Zn)
Lime (As, Pb)
Pesticides (As, Hg, Pb, Cu, Mn)
Irrigation water (As in West Bengal, Bangladesh, Pb in lead mining area, Se, Cr in leather industrial area)
Waste disposal on land
Sewage and sludge (Cd, Cr, Ni, Zn, Cu, Pb, Hg, Fe)
Leachate from landfill (As, Cd, Fe, Pb, Zn)
Scrapheaps (Cd, Cr, Cu, Pb, Zn)
Coal ash (Cu, Pb, Cr, Cr, Ni, Fe)

soil environment (Dotaniya et al. 2017a). Increasing moisture also enhances the mobility kinetics in soil and more change to damage the plant growth (Eick et al. 1999). Faulty crop input application strategies are also enhanced the significant amounts of Pb in surface layers (de Abreu et al. 1998).

Pb concentration in soil reduces healthy growth of plants and causes different ill effect, such as poor growth, change in water balance, yellowing and chlorosis, blackening of plant roots, imbalance in mineral nutrition, poor cell wall growth, hormonal imbalance, and reduction in root exudation process and thus mediating cell permeability and structure. It affects the protoplasmic reaction and process and ultimately stunted growth or dead of plant. In soil, the common adverse mechanisms of Pb include reducing the plant nutrient availability kinetics, mineralization of OM, soil microbial population and diversity, and poor soil structure. Increasing the Pb availability in soil solution reduces the availability of other metal cations and plants show deficiency symptoms (Kim et al. 2002). Increasing the Pb levels reduced the Ni concentration in spinach under Vertisol of Madhya Pradesh, India (Dotaniya and Pipalde 2018). A similar pattern of elevated level of Pb in black soil adversely affects the soil enzymatic activates and maximum reduction had with DHA than alkaline and acid phosphatases (Pipalde and Dotaniya 2018). Stefanov et al. (1995) reported that long-term (41 years) application of Pb containing fertilizers accumulated significant amounts of available Pb in soil. Other scientific reports showed that use of mine industry effluent accumulated significant amounts of Pb and other heavy metals in sediment particles (Laxen and Harrison 1977). In this chapter, we have discussed the Pb toxicity in soil–plant system and sustainable management for crop production.

2   Sources of Lead Contamination

Today, lead contamination in environment, soil and food chain is increasing at a faster pace and has led to contamination of natural ecosystems. Lesser concentration for longer exposure may lead to toxic levels and interfere with the ecosystem services and reduce the system productivity. Soil–plant system Pb mediated plant nutrient dynamics and the uptake kinetic in plants. It adversely affects the plant metabolism and ultimately reduces the crop growth and yield. These situations are more pathetic in vegetable production areas mainly with the only source of Pb containing wastewater (WW). Several research reports are the witness of the Pb levels increasing with time in food chain either terrestrial or aquatic food web (Fig. 1).

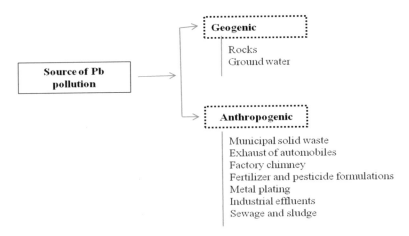

Fig. 1 Sources of Pb pollution in environment (Modified from Dotaniya et al. 2018c)

2.1 Lead Concentration in Atmosphere

Atmosphere is a larger sink for the Pb emission and contributing mainly by the petrol emission and significant chunk by the industrial activities. These contributions are greatly decided by the type of industry and concentration of Pb after primary treatment at industry level, population and lifestyle, traffic density as well as weather conditions of particular areas. Metropolitan cities are the big contributor of Pb concentration in air causing lungs problems. The international organization studied on the Pb concentration in air with respect to petroleum type, distance from road, vehicle type and green belt of the trees (Gloag 1981). Increasing the petrol efficiency at industrial level addicting the tetraethyl and tetramethyl Pb are the main cause of Pb emission through vehicles (Sharma and Dubey 2005). The concentration of Pb inversely reduces with increasing the distance from road. However, the dense green belt of trees significantly reduces the Pb concentration in air and cleans the environment. Road side plant leaves contain more concentration than the plants away from the dense traffic road. The mean concentration of Pb in air is low as 0.02 μg m^{-3} vary in remote locations but higher concentration was observed in European Community metropolitan urban areas up to 10 μg m^{-3} (World Health Organisation 1977).

2.2 Sewage Channels

Most of the developing countries are facing the challenges sustainable food production with quality aspect. Natural resources are indiscriminating exploiting to enhance the system productivity without rational management (Dotaniya et al. 2016a). Use of sewage water for agricultural crop production system is a common practice in water scare area of peri urban (Dotaniya et al. 2018a). It contains meager amount of Pb due to intermixing of household and smaller industrial effluent. Long-term application of these marginal quality water accumulated significant amounts of Pb in soil. Application of Pb at $0–100$ mg kg^{-1} reduces the soil enzymatic activities and adversely affected the plant nutrient supply chain towards crop plants (Dotaniya and Pipalde 2018).

2.3 Through Contaminated Food

Food chain contamination is the main source of Pb contamination in human body. Crop cultivation on marginal or polluted natural resources produces the metal contaminated food stuffs and absorb by different organs during the intake. These situations are very pathetic in urban polluted areas and peoples are forced to use contaminated food materials produce from industrial waste, sewage effluent, marginal land and spurious crop input materials. Leafy crops absorb more amount of Pb from polluted soils as compare to grain crops (Dotaniya et al. 2018b). The Pb metal toxicity in human being can be calculated by using Pierzynski et al. (2000) hazard quotients (HQgv). Dotaniya et al. (2018c) suggested that application of organic substances like crop residue, FYM, vermicomposting reduce the transfer of Pb from soil to edible plant parts. Similar pattern inorganic substances are also reduce the mobility and availability of Pb in soils.

2.4 Lead in Water

Water is also a prime source of Pb in human beings and affects various biochemical processes. The geogenic concentration of Pb causes groundwater pollution and less chance to minimize the toxicity across the globe. Across the world soil Cr, Cu, Ni and Zn concentration are 68, 22, 22 and 66, respectively (Das and Chakrapani 2011). These values might be different due to regional and climatic variations. Few Pb pollution hot spots are well known and different agencies are working to reduce the concentration in drinking water. However, long-term application of sewage or industrial effluent also contaminated the ground water and pollute the drown streams of water. Dumping of industrial waste without any treatment on healthy soil or in freshwater bodies also accelerated the groundwater pollution.

2.5 Lead in Soil

The geogenic contribution of Pb is the main source of Pb contamination. Industrial use explores the extraction of Pb mineral for human use and spread the concentration in remote areas. These processes are speedier after the industrial revolutions and liberalization period across the globe. The concentration in soil reduces the soil microbial activity, population and also diversity.

3 Lead Toxicity in Plants

Lead is one of the toxic pollutants and geogenic evolution. Due to fast global industrial development enhance the concentration of Pb in burgeoning crop areas (Nriagu 1996). The available concentration of Pb in soil reached plant parts via root uptake. Dark green leaves, stunted growth, chlorosis and plant look like infested by insect-pest are the important symptoms of Pb toxicity in plants (Jones et al. 1973). Initially Pb reaches in root apoplast and after passing few processes reach endodermis through cortex through passive uptake mechanism. Here, it would be accumulated significant amounts and actively involve as a barrier to transfer Pb from root to shoot. This might be one of the potential clues that root biomass having higher content of Pb than shoot parts (Verma and Dubey 2003). Higher concentration of Pb in soil moves in other parts like vascular tissues and diffuses out in nearby tissues. These evidences showed the movement of Pb into the symplast. The carboxyl groups of galacturonic acid act as a binding agent in cell wall and reduce the transportation mechanism through the apoplast in most of the plants (Rudakova et al. 1988). In lower concentration, soils predominantly move the Pb ions into the apoplast but toxic level it damage barrier tissue and huge concentration entered into cells (Sharma and Dubey 2005). Higher proportion of readily available Pb in soil solution promotes more amount of Pb in shoot via root cells. The update pattern is very much dependent on genetic potential of plant, age and environmental factors (Dotaniya et al. 2018c). More leafy crops are having higher concentration of Pb in root and shoot as compare to other grain crops. Those crops having higher amount of lignified tissues is acted Pb sink than less lignified (Gloag 1981). Lane and Martin (1977) suggested that testa prevent internal entry of Pb in the cell till the ruptured by the developing radicle, after that available Pb is movement in the cell. Huang and Cunningham (1996) reported that dicotyleden plants accumulated more concentration in plant parts than monocot crops. The absorb concentration of Pb from soil was accumulated as follows root > leaf > stem > flower > seed parts (Antosiewicz 1992). These are the standard sequence of Pb accumulation, whereas it may vary with plant species and plant age. Later on, Godzik (1993) reported that older leaves are having higher amount of absorb Pb than newly leaves. The plants are having higher intercellular space, cell wall size; lignified tissues, vacuole number and size are accumulated major amount and minor concentration in the endoplasmic reticulum, dictyosomes and associated tissues. According to the published research paper approximately 96 percent of total absorbed Pb from soil solution is deposited only in cell wall and vacuole (Sharma and Dubey 2005). Thick cells act as a strong barrier to Pb transportation in plant parts from soil and in toxic conditions accumulate larger amounts of it. Significant amounts of Pb in plant parts damage the cell and causes an imbalance in the metal barrier system of plasmalemma as well as tonoplast (Seregin et al. 2004). Some of the metabolic activities affected by the concentration of Pb in plants are listed in Table 2.

The Pb persistence in environment depends on the availability of Pb, organic matter in soil, pH and climatic factors (Punamiya et al. 2010). In most of the soil, Pb exists in free ions, complexes with organic acids or inorganic substances like carbonate and bicarbonate, sulphate and chloride ions. In higher percent of clay containing soil are also having higher potential to adsorb the Pb and reduce the intensity. Across the global research findings showed that increasing humus content in soil reduce the Pb toxicity. For example, sewage having significant amounts of organic matter and long- term application accumulated huge amount of C in soil, which is formed humus-metal complex and reduce the available toxic form of Pb towards plants. Long- term application of Pb containing industrial waste, accumulated significant amounts of Pb on surface layer due to higher amount of organic carbon (Kopittke et al. 2008). The presence of other cationic plant nutrients in soil solution reduces the Pb uptake by plant roots due to similar uptake channel. Application of Ca containing fertilizers reduces the Pb uptake by crop plants (Kim et al. 2002). Use

Table 2 Effect of Pb concentration on enzymatic activities of plant metabolic process

Metabolic process	Associated enzyme	Crop plant	Effect	References
Chlorophyll synthesis	δ-aminolaevulinate	*Pennisetum typhoideum*	Inhibition	Prasad and Prasad (1987)
CO₂ fixation	Ribulose-1,5, bis phosphate	*Avena sativa*	Inhibition	Moustakas et al. (1994)
	Phosphoenol pyruvate carboxylase	*Zea mays*	Inhibition	Vojtechova and Leblova (1991)
Calvin	Glyceraldehyde 3-phosphate dehydrogenase	*Spinach oleracea*	Inhibition	Vallee and Ulmer (1972)
	Ribulose 5-phosphate kinase	*Spinach oleracea*	Inhibition	Vallee and Ulmer (1972)
Pentose phosphate pathways	Glucose 6-phosphate dehydrogenase	*Spinach oleracea*	Inhibition	Vallee and Ulmer (1972)
N₂ assimilation	Nitrate reductase	*Cucumis sativus*	Inhibition	Burzynski (1987)
	Glutamine synthetase	*Glycine max*	Inhibition	Lee et al. (1976)
Nucleolytic enzymes	Deoxyribonuclease	*Hydrilla verticillata*	Increase	Jana and Choudhari (1982)
	Ribonuclease	*Hydrilla verticillata*	Increase	Jana and Choudhari (1982)
Protein synthesis	Protease	*Hydrilla verticillata*	Increase	Jana and Choudhari (1982)
Phosphohydrolase	Alkaline phosphatase	*Hydrilla verticillata*	Increase	Jana and Choudhari (1982)
	Acid phosphatase	*Glycine max*	Increase	Lee et al. (1976)
Sugar metabolism	A-amylase	*Oryza sativa*	Inhibition	Mukherji and Maitra (1976)
Energy generation	ATP synthetase	*Zea mays*	Inhibition	Tu Shu and Brouillette (1987)
	ATPase	*Zea mays*	Inhibition	Tu Shu and Brouillette (1987)
Antioxidative metabolism	Catalase	*Oryza sativa*	Inhibition	Verma and Dubey (2003)
	Guaiacol peroxidase	*Glycine max*	Increase	Lee et al. (1976)
	Ascorbate oxidase	*Phaseolus aureus*	Increase	Rashid and Mukherji (1991)
	Ascorbate peroxidase	*Oryza sativa*	Increase	Verma and Dubey (2003)
	Glutathione reductase	*Oryza sativa*	Increase	Verma and Dubey (2003)
	Superoxide dismutase	*Oryza sativa*	Increase	Verma and Dubey (2003)

of plants (phytoremediation) is to remediate the Pb concentration from soil-plant systems. These plants are having higher capacity of Pb uptake without affecting plant metabolism and crop yield. These accumulated larger amounts of Pb in plant cell vacuoles as waste materials. Some of the parameters/ratios, which indicated whether a plant is phytoremediation or not are as follows:

1. Bioconcentration factor (BCF) defines the contamination removal capacity of the plant and was calculated by the given formula (Zhuang et al. 2007).

$$BCF = \frac{Pb_{harvested\ tissue}}{Pb_{soil}}$$

where, $Pb_{harvested\ tissue}$ is a concentration of Pb in harvested plant parts (root, shoot); and Pb_{soil} is concentration of Cr in soil of respective treatment. If this value greater than 1 means crop plant behave like hyperaccumulator.

2. Translocation factor (TF) means transfer of Pb metal ions from root to shoot and quantified by the formula proposed by Adesodun et al. (2010).

$$TF = \frac{Pb_{shoot}}{Pb_{root}}$$

where, Pb_{shoot} and Pb_{root} are concentration of Cr in root and shoot, respectively.

3. Translocation efficiency (TE) was calculated from the formula described by Meers et al. (2004) as below.

$$TE\,(\%) = \frac{Pb_{content\ in\ shoot}}{Pb_{content\ in\ whole\ plant}} X\,100$$

4. Lead removal represent the Pb removal capacity of the crop and was calculated as per given formula.

$$Pb\,removal\,(\%) = \frac{Total\,Pb\,uptake\,by\,plant}{Total\,Pb\,applied\,to\,soil} X\,100$$

4 Lead Toxicity Effect on Ecosystem Services

Ecosystem services are the supporting wheel of the living creates on the earth. These are interlinking each other and can be easily computed the effect of a factor with respect to the health of the ecosystem. Lead concentration significantly affects the ecosystem potential productivity and its health (Fig. 2). Some of the ecosystem services are enlisted below:

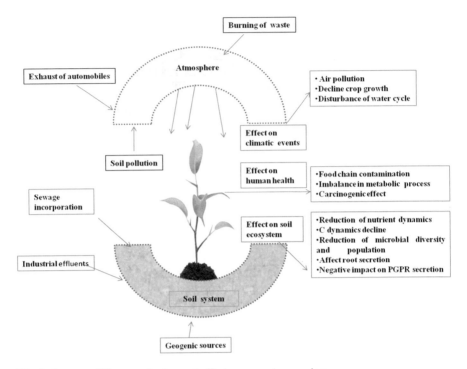

Fig. 2 Sources of Pb contamination and effect on ecosystem services

4.1 Soil Forming Process

Pedogenic processes are the key to formation of soil. Important chemical and bio-geochemical processes that occurs in the soils are mineralization-immobilization, dissolution-precipitation, sorption-desorption and oxidation-reduction (Sanyal and Majumdar 2009). They are involved in influencing dynamics of several nutrients, soil carbon dynamics and ultimately soil biodiversity by creating nutrition competition. These processes are affected by many factors and population and type of organisms are one of them. Increasing the concentration of Pb is unbalanced the microorganism population by mediating the availability of food material during mineralization and more availability of biotoxic Pb. Increasing the concentration of Pb from 0 to 100 mg kg^{-1} reduces the soil enzymatic activities (dehydrogenase activity, acid and phosphatase activities). Among the enzymatic activities DHA had declined more as compared to other two soil enzymes (Dotaniya and Pipalde 2018). These enzymes are very much important for the availability of plant nutrients in soil and formation of good soil.

4.2   Plant Nutrient Dynamics

Plant nutrients are essential for growth and sustainable crop yield (Beri et al. 1995). Toxic level of Pb in soil reduces the plant nutrients and plant suffered with abiotic stress. In common, increasing level of Pb reduce the uptake of cations like K^+, Ca^{2+}, Mg^{2+}, micronutrients (Zn, Cu, Ni, Fe) as well as the uptake of N from soils (Godbold and Kettner 1991; Yongsheng et al. 2011). During the uptake, it affects the mechanism of uptake and blocks the plant nutrient uptake towards plant parts via root (Sharma and Dubey 2005). Addition of plant derived residue enhances the plant nutrient concentration in soil solution and reduces the metal toxicity (Dotaniya et al. 2016a).

4.3   Soil Microbial Count

Directly and indirectly Pb toxicity reduces the microbial count in soil-plant environment. Soil organic matter decomposition rate was declined due to present of metal concentration in soil (Bahar et al. 2012). The biotoxic effect reduces the diversity of soil microorganisms, richness and population (Akerblom et al. 2007; Banat et al. 2010). Application of Pb @ 100 mg kg^{-1} reduces the carbon mineralization rate and soil enzymatic activities. Many research findings are showed that heavy metal toxicity reduces the soil health and plant growth. Higher concentration of Pb in soil reduces the population of *Rhizobia* and severely affects the biological N fixation potential of crops (Hernandez et al. 2003).

4.4   Plant Secretion

Plant nutrient dynamics are the most important factor for plant growth. Healthy soils produced good amount of plant biomass and yield. During the respiration or other metabolic process, plants secrete significant amounts of low-molecular-weight organic acids (LMWOAs) in soil. Most of the chemoautotrophic soil microorganisms are taking food from them and enhance the nutrient mineralization process. Dotaniya et al. (2019) described that 40–70% of photosynthetic amount released as root exudates. These are amino acids like acetic, oxalic, and citric acids; also plant growth hormones. Significant amounts of Pb in soil disturb the plant root exudation process and reduce the secretion of LMWOAs in soil due to poor growth of crop plants. These situations are very critical in high contamination level and soils are look like barren land and not to produce even bunch of grasses. Lead is not an oxido-reducing metal and produces oxidative stress on plant and generated ROS, which are acted as a phytotoxicity effect on plant growth (Pinto et al. 2003).

5   Factors Affecting Lead Mobility in Soil–Plant System

Heavy metal dynamics are affected by the soil, management and climatic factors. Most of the cases soil properties like pH, clay content, cation exchange capacity, soil structure, organic matter and amount of other metal ions in the soil are decided the quantum of metal toxicity towards plants (Meena et al. 2019b). Higher clay content reduces the availability of Pb by forming clay-humus complex. For that addition of organic matter reduce the mobility and toxicity of the heavy metals (Dotaniya et al. 2016b). These are also provided the C to soil microorganisms as a food material and enhance the microbial growth and diversity. The carboxylic group of organic matter bound the metal ions and reduces the metal availability. Most of the cationic heavy metals including Pb are having higher mobility in acid environment (Table 3).

Introducing acid neutralizing material in soil by inorganic substances reduces the Pb mobility in soil and uptake by crop plants. Water stress also reduced the heavy metal mobility and toxicity in soils (Dotaniya et al. 2018b). The Pb forms in soil are also important to assess the toxicity; oxyanion complexes are the most available form of Pb in soil and groundwater streams (Dotaniya et al. 2018c). The higher application of phosphatic fertilizers reduces the Pb toxicity and improves the plant growth in contaminated soils (Lenka et al. 2016). Dotaniya and Pipalde (2018) reported that increasing the concentration of Ni in black soil reduce the Pb toxicity up to 100 mg kg^{-1} level (Table 4). Both are cationic in nature and antagonistic effect

Table 3 Heavy metal mobility in soil (Adopted from Ferguson 1990)

Mobility	Oxidizing	Acid	Neutral (alkaline)	Reducing
Very high	–	–	Se	–
High	Se	Se, Hg	–	–
Medium	Hg, As, Cd	As, Cd	As, Cd	Ti
Low	Pb, As, Sb, Ti	Pb, Bi, Sb, Ti	Pb, Bi, Sb, Ti, In	–
Very low to immobile	Te	Te	Te, Hg	Te, Se, Hg, As, Cd, Pb, Bi, Ti

Table 4 Effect of Pb and Ni levels on Ni concentration (mg kg^{-1}) in shoot

Pb (mg kg^{-1})	Ni (mg kg^{-1})			
	100	150	300	Mean
0	7.033	10.400	13.900	10.444
100	3.967	4.600	11.200	6.589
150	3.400	3.433	7.567	4.800
300	3.067	3.900	7.533	4.833
Mean	4.367	5.583	10.050	
LSD ($p = 0.05$)	Pb = 0.510; Ni = 0.510; Pb X Ni = 1.020			

reduces the toxicity of Pb, however increasing the Ni levels enhance the growth to reduce the adverse effect of Pb (Pipalde and Dotaniya 2018).

Dynamics of Pb in plant system are also affected by the plant species. Some plants in nature are having excluder strategies to reduce the uptake of Pb in plant parts through roots (Baker 1981). Other type of plant is having the higher potential to accumulate the Pb concentration without affecting the crop yield. These plants are known as hyperaccumulator and used for phytoremediation mechanisms. Plant is also involved to follow the path avoidance, detoxification and biochemical tolerance to minimize the abiotic stress in plant (Dotaniya et al. 2018d). Losses of chlorophyll content and disturbance of respiration and photosynthesis rate are the key symptoms of Pb toxicity in plant. Lead toxicity is caused much harmful effect in younger plant compare to older plants. Leafy crops accumulated more amount of Pb than grain crops. Peri-urban vegetable production system is having more chance to Pb contamination. Food chain contamination of Pb caused carcinogenic effect in human beings. Infant are more sensitive to Pb toxicity because of stunted growth, poor mental development, hair fall and failure of organ and metabolic process. Different metal ratios (bioconcentration factor, transfer factor, translocation efficiency, metal removal, geo-accumulation indices) are helpful to assess the phyto-accumulation potential of a crop plant (Dotaniya et al. 2017b; Dotaniya et al. 2019).

6 Conclusions

Heavy metals are biotoxic in nature and its toxicity causes imbalance in ecosystem functions. Among heavy metals, Pb plays an important role and is categorized as a potential hazardous metal. Growing anthropogenic activities gained huge attention to prevent its concentration in food chain contamination by use of organic and inorganic substances. A meager amount of Pb in soil affects the soil microbial population and diversity and reduces the plant nutrient dynamics and crop yield. Environment impact analysis should be done at certain period of time with respect to ecosystem services. Developing high-efficiency phytoremediation varieties to reduce Pb concentration from soil system is essential. Regular monitoring of industrial effluent channels as per the WHO guidelines and discharging effluent in field crops after treatment are needed to reduce Pb toxicity in ecosystem. Hence, research priorities should be directed towards Pb mobility in soil–plant–human continuum. Periodical monitoring of foodstuffs from production unit to the finally consumption level of the food chain for formulation of preventive measures has to be done.

Acknowledgements The authors earnestly thank Dr M. D. Meena, Scientist, ICAR-DRMR, Bharatpur, India for the help during the writing of the manuscript.

References

de Abreu CA, de Abreu MF, de Andrade JC (1998) Distribution of lead in the soil profile evaluated by DTPA and Mehlich-3 solutions. Bragantia 57:185–192

Adesodun JK, Atayese MO, Agbaje TA, Osadiaye BA, Mafe OF, Soretire AA (2010) Phytoremediation potentials of sunflowers (*Tithonia diversifolia* and *Helianthus annus*) for metals in soils contaminated with zinc and lead nitrates. Water Air Soil Pollut 207:195–201

Akerblom S, Baath E, Bringmark L, Bringmark E (2007) Experimentally induced effects of heavy metals on microbial activity and community structure of forest MOR layers. Biol Fertil Soils 44:79–91

Aktar MW, Sengupta D, Chowdhury (2009) Impact of pesticides use in agriculture: their benefits and hazards. Interdiscip Toxicol 2:1–12

Alloway BJ (1990) Heavy metals in soils. Wiley, New York

Antosiewicz DM (1992) Adaption of plants to an environment polluted with heavy metals. Acta Soc Bot Polon 61:281–299

Bahar MM, Megharaj M, Naidu R (2012) Arsenic bioremediation potential of a new arsenite-oxidizing bacterium *Stenotrophomonas* sp. MM-7 isolated from soil. Biodegradation 23:803–812

Baker AJM (1981) Accumulators and excluders-strategies in the response of plant to heavy metals. J Plant Nutr 3:326–329

Banat IM, Franzetti A, Gandolfi I, Bestetti G, Martinotti MG, Fracchia L, Marchant R (2010) Microbial biosurfactants production, applications and future potential. Appl Microbiol Biotechnol 87:427–444

Bell D, MacLeod AF (1983) Dieldrin pollution of a human food chain. Human Toxic 2:75–82

Beri V, Sidhu BS, Bahl GS, Bhat AK (1995) Nitrogen and phosphorus transformations as affected by crop residue management practices and their influence on crop yield. Soil Use Manage 11:51–54

Bharti VS, Dotaniya ML, Shukla SP, Yadav VK (2017) Managing soil fertility through microbes: prospects, challenges and future strategies. In: Singh JS, Seneviratne G (eds) Agro-environmental sustainability. Springer, p. 81–111

Burzynski M (1987) The influence of lead and cadmium on the adsorption and distribution of potassium, calcium, magnesium and iron in cucumber seedlings. Acta Physiol Plant 9:229–238

Das SK, Chakrapani GJ (2011) Assessment of trace metal toxicity in soils of Raniganj Coalfield, India. Environ Monit Assess 177:63–71

Dotaniya ML, Pipalde JS (2018) Soil enzymatic activities as influenced by lead and nickel concentrations in a Vertisol of Central India. Bull Environ Contam Toxicol 101:380–385

Dotaniya ML, Datta SC, Biswas DR, Dotaniya CK, Meena BL, Rajendiran S, Regar KL, Lata M (2016a) Use of sugarcane industrial byproducts for improving sugarcane productivity and soil health-a review. Intl J Recyc Org Waste Agric 5:185–194

Dotaniya ML, Meena VD, Kumar K, Meena BP, Jat SL, Lata M, Ram A, Dotaniya CK, Chari MS (2016b) Impact of biosolids on agriculture and biodiversity. Today and Tomorrow's Printer and Publisher, New Delhi, p. 11–20

Dotaniya ML, Rajendiran S, Coumar MV, Meena VD, Saha JK, Kundu S, Kumar A, Patra AK (2017a) Interactive effect of cadmium and zinc on chromium uptake in spinach grown on Vertisol of Central India. Intl J Environ Sci Tech 15:441–448

Dotaniya ML, Meena VD, Rajendiran S, Coumar MV, Saha JK, Kundu S, Patra AK (2017b) Geo-accumulation indices of heavy metals in soil and groundwater of Kanpur, India under long term irrigation of tannery effluent. Bull Environ Contam Toxicol 98:706–711

Dotaniya ML, Meena VD, Rajendiran S, Coumar MV, Sahu A, Saha JK, Kundu S, Das H, Patra AK (2018a) Impact of long-term application of Patranala sewage on carbon sequestration and heavy metal accumulation in soils. J Indian Soc Soil Sci 66:310–317

Dotaniya ML, Rajendiran S, Meena VD, Coumar MV, Saha JK, Kundu S, Patra AK (2018b) Impact of long-term application of sewage on soil and crop quality in Vertisols of central India. Bull Environ Contam Toxicol 101:779–786

Dotaniya ML, Panwar NR, Meena VD, Dotaniya CK, Regar KL, Lata M, Saha JK (2018c) Bioremediation of metal contaminated soils for sustainable crop production. In: Meena VS (ed) Role of rhizospheric microbes in soil. Springer, p. 143–173

Dotaniya ML, Rajendiran S, Dotaniya CK, Solanki P, Meena VD, Saha JK, Patra AK (2018d) Microbial assisted phytoremediation for heavy metal contaminated soils. In: Kumar V, Kumar M, Prasad R (eds) Phytobiont and ecosystem restitution. Springer, p. 295–317

Dotaniya ML, Meena VD, Saha JK, Rajendiran S, Patra AK, Dotaniya CK, Meena HM, Kumar K, Meena BP (2019) Environmental impact measurements: tool and techniques. In: Martínez L, Kharissova O, Kharisov B (eds) Handbook of ecomaterials. Springer, p. 1-31.

Eick MJ, Peak JD, Brady PV, Pesek JD (1999) Kinetics of lead adsorption and desorption on goethite: Residence time effect. Soil Sci 164:28–39

Ferguson JE (1990) The heavy elements: chemistry, environmental impact and health effects. Pergamon Press, New York

Gloag D (1981) Sources of lead pollution. Br Med J 282:41–44

Godbold DL, Kettner C (1991) Lead influence on root growth and mineral nutrition of *Piceaabies* seedlings. J Plant Physiol 139:95–99

Godzik B (1993) Heavy metal contents in plants from zinc dumps and reference area. Pol Bot Stud 5:113–132

Hernandez L, Probst A, Probst JL, Ulrich E (2003) Heavy metal distribution in some French forest soils: evidence for atmospheric contamination. Sci Total Environ 312:195–219

Huang JW, Cunningham SD (1996) Lead phytoextraction: species variation in lead uptake and translocation. New Phytol 134:75–84

Jana S, Choudhari MA (1982) Senescence in submerged aquatic angiosperms: effect of heavy metals. New Phytol 90:477–484

Jones LHP, Clement CR, Hopper MJ (1973) Lead uptake from solution by perennial ryegrass and its transport from root to shoots. Plant Soil 38:403–414

Kim YY, Yang YY, Lee Y (2002) Pb and Cd uptake in rice roots. Physiol Plant 116:368–372

Kopittke PM, Asher CJ, Kopittke RA, Menzies NW (2008) Prediction of Pb speciation in concentrated and dilute nutrient solutions. Environ Pollut 153:548–554

Lane SD, Martin ES (1977) A historical investigation of lead uptake in *Raphanussativus*. New Phytol 79:281–286

Laxen DPH, Harrison RM (1977) The highway as a source of water pollution: an appraisal of heavy metal lead. Water Res 11:1–11

Lee KC, Cunningham BA, Poulsen GM, Liang JM, Moore RB (1976) Effect of cadmium on respiration rate and activities of several enzymes in soybean seedlings. Physiol Plant 36:4–6

Lenka S, Rajendiran S, Coumar MV, Dotaniya ML, Saha JK (2016) Impacts of fertilizers use on environmental quality. In: National Seminar on Environmental concern for fertilizer use in future at Bidhan Chandra Krishi Viswavidyalaya, Kalyani, India on February 26, 2016

Levinson AA (1974) Introduction to exploration geochemistry. Applied Publishing, Calgary

Meena VD, Dotaniya ML, Saha JK, Patra AK (2015) Antibiotics and antibiotic resistant bacteria in wastewater: impact on environment, soil microbial activity and human health. Afr J Microbiol Res 9:965–978

Meena MD, Yadav RK, Narjary B, Yadav G, Jat HS, Sheoran P, Meena MK, Antil RS, Meena BL, Singh HV, Singh Meena V, Rai PK, Ghosh A, Moharana PC (2019a) Municipal solid waste (MSW): Strategies to improve salt affected soil sustainability: a review. Waste Manag 84:38–53

Meena VD, Dotaniya ML, Saha JK, Meena BP, Das H, Beena, Patra AK (2019b) Sustainable C and N management under metal-contaminated soils. In: Datta R, Meena RS, Pathan S, Ceccherini M (eds) Carbon and nitrogen cycling in soil. Springer, p. 293–336

Meers E, Hopgood M, Lesage E, Vervaeke P, Tack FMG, Verloo M (2004) Enhanced phytoextraction: in search for EDTA alternatives. Int J Phytorem 6:95–109

Moustakas M, Lanaras T, Symeonidis L, Karataglis S (1994) Growth and some phytosynthetic characteristics of field grown *Avena sativa* under copper and lead stress. Phytosynthetica 30:389–396

Mukherji S, Maitra P (1976) Toxic effect of lead on growth and metabolism of germinating rice root tip cells. Ind J Exp Biol 14:519–521

Nriagu JO (1996) History of global metal pollution. Sci 272:223–224

Pierzynski GM, Sims JT, Vance GF (2000) Soils and Environmental Quality. FL. CRC Press, Boca Raton, p. 155–207

Pinto E, Sigaud-Kutner TCS, Leitao AS, Okamoto OK, Morse D, Coilepicolo P (2003) Heavy metal induced oxidative stress in algae. J Phycol 39:1008–1018

Pipalde JS, Dotaniya ML (2018) Interactive effects of lead and nickel contamination on nickel mobility dynamics in spinach. Intl J Environ Res 12:553–560

Prasad DDK, Prasad ARK (1987) Altered δ-aminolae-vulinic acid metabolism by lead and mercury in germinating seedlings of Bajra (*Pennisetum typhoideum*). J Plant Physiol 127:241–249

Punamiya P, Datta R, Sarkar D, Barber S, Patel M, Das P (2010) Symbiotic role of glomus mosseae in phytoextraction of lead in vetiver grass [*Chrysopogon zizanioides* (L.)]. J Hazard Mater 177:465–474

Rajendiran S, Dotaniya ML, Coumar MV, Panwar NR, Saha JK (2015) Heavy metal polluted soils in India: status and countermeasures. JNKVV Res J 49: 320–337

Rajendiran S, Singh TB, Saha JK, Coumar JK, Dotaniya ML, Kundu S, Patra AK (2018) Spatial distribution and baseline concentration of heavy metals in swell–shrink soils of Madhya Pradesh, India. In: Singh V, Yadav S, Yadava R (eds) Environmental pollution. Water science and technology Library. Springer, p. 135–145

Rashid P, Mukherji S (1991) Change in catalase and ascorbic oxidase activities in response to lead nitrate treatments in mungbean. Indian J Plant Physiol 34:143–146

Rudakova EV, Karakis KD, Sidorshina ET (1988) The role of plant cell walls in the uptake and accumulation of metal ions. Fiziol Biochim Kult Rast 20:3–12

Saha JK, Rajendiran S, Coumar MV, Dotaniya ML, Kundu S, Patra AK (2017) Soil pollution—an emerging threat to agriculture. Springer

Sanyal SK, Majumdar K (2009) Nutrient dynamics in soil. J Indian Soc Soil Sci 57:477–493

Seregin IV, Shpigun LK, Ivaniov VB (2004) Distribution and toxic effect of cadmium and lead on maize roots. Russ J Plant Physiol 51:525–533

Sharma P, Dubey RS (2005) Lead toxicity in plants. Braz J Plant Physiol 17:35–52

Solanki P, Narayan M, Meena SS, Srivastava RK, Dotaniya ML, Dotaniya CK (2018) Phytobionts of wastewater and restitution. In: Kumar V, Kumar M, Prasad R (eds) Phytobiont and ecosystem restitution. Springer, p. 379–401

Stefanov K, Seizova K, Popova I, Petkov VL, Kimenov G, Popov S (1995) Effect of lead ions on the phospholipids composition in leaves of *Zea mays* and *Phaseolus vulgaris*. J Plant Physiol 147:243–246

Tu Shu I, Brouillette JN (1987) Metal ion inhibition of corn root plasmamembrane ATPase. Metal ion inhibition of corn rot plasma membrane ATPase. Phytochemistry 26:65–69

Vallee BL, Ulmer DD (1972) Biochemical effects of mercury, cadmium and lead. Annu Rev Biochem 41:91–128

Verma S, Dubey RS (2003) Lead toxicity induces lipid peroxidation and alters the activities of antioxidant enzymes in growing rice plants. Plant Sci 164:645–655

Vojtechova M, Leblova S (1991) Uptake of lead and cadmium by maize seedlings and the effect of heavy metals on the activity of phosphoenolpyruvate carboxylase isolated from maize. Biol Planta 33:386–394

World Health Organisation (1977) Environmental health criteria No. 52, Chapter 3. Lead

Yongsheng W, Qihui L, Qian T (2011) Effect of Pb on growth, accumulation and quality component of tea plant. Procedia Engineer 18:214–219

Zhuang P, Yang QW, Wang HB, Shu WS (2007) Phytoextraction of heavy metals by eight plant species in the field. Water Air Soil Pollut 184:235–242

Lead Toxicity in Plants: A Review

Anindita Mitra, Soumya Chatterjee, Anna V. Voronina, Clemens Walther, and Dharmendra K. Gupta

Abstract The harmful effects of lead (Pb) contamination are well known. Accumulation of Pb in the soil due to natural and anthropogenic sources causes substantial problems to soil biota and the milieu, which is of immense concern to the scientific community. Both stable and isotopic Pb, which is naturally present in the environment, can be accumulated within vegetation also disturbing plant growth and food safety. Many plants have developed detoxification mechanisms to accumulate Pb within their body without any harmful effects. Phytoremediation practices may be the better option for amelioration of Pb contaminated soil. This chapter deals with the recent advancements in the field of lead contamination and remediation through bioremediation for safe use of lead-contaminated areas.

Keywords Plants · Remediation · Phytoremediation · Genotoxicity

A. Mitra
Department of Zoology, Bankura Christian College, Bankura, West Bengal, India

S. Chatterjee
Defence Research Laboratory, DRDO, Tezpur, Assam, India

A. V. Voronina
Radiochemistry and Applied Ecology Department, Ural Federal University, Physical Technology Institute, Ekaterinburg, Russia

C. Walther
Gottfried Wilhelm Leibniz Universität Hannover, Institut für Radioökologie und Strahlenschutz (IRS), Hannover, Germany

D. K. Gupta (✉)
Ministry of Environment, Forest and Climate Change, Indira Paryavaran Bhavan, Jorbagh Road, Aliganj, New Delhi, India
e-mail: guptadk1971@gmail.com

© Springer Nature Switzerland AG 2020
D. K. Gupta et al. (eds.), *Lead in Plants and the Environment*, Radionuclides and Heavy Metals in the Environment,
https://doi.org/10.1007/978-3-030-21638-2_6

1 Introduction

Lead (Pb) is an industry-friendly heavy metal due to its selective properties like high density, lower melting point, acid resistance and easy moulding. Due to its versatile anthropogenic utilities, the metal is also responsible to pollute environment in a varied degree. Soil contamination with Pb is a considerable concern now-a-days as it can affect directly all living beings (Arshad et al. 2008; Ma et al. 2016). Therefore, receive much scientific attention and studied in a comprehensive way. Like other heavy metals (HMs), Pb also enters into the agroecosystem through contaminated soil and enters into the food chain (Gupta and Li 2013; Anjum et al. 2016; Ashraf et al. 2017). Pb accumulation in the plants poses a considerably grave health risk for humans. In terrestrial ecosystem, soil serves as chief source of HMs for plants, which after getting entry within plant system, lead to major consequences in crop production and grain quality. European REACH regulations labelled Pb as "the chemical of great concern" and categorized it as the second most harmful pollutant after arsenic (Pourrut et al. 2011a). Lead directly or indirectly affect photosynthesis, plant's nutrient uptake, seedling growth, enzyme activities, water imbalance, and membrane permeability (Sharma and Dubey 2005; Shahid et al. 2011; Kumar et al. 2012). At a higher concentration, Pb interfere with the chloroplasts normal functioning by preventing the enzymes involved in chlorophyll biosynthesis, CO_2 fixation, and the pigment-protein complexation of photosystems (Sharma and Dubey 2005). Photosystem II (PSII) is more prone to Pb exposure than that of PSI (Romanowska et al. 2012) as it is collapsed in a higher extent including detrimental effects on both the donor and acceptor sites, oxygen evolving complex and electron-transfer reactions (Pourrut et al. 2011a). Most of the HMs including Pb induce the higher production of ROS (at toxic level) within plant cell and subsequently augment the processes like, peroxidation of lipids, proportion of saturated fatty acids versus unsaturated fatty acids (which increase in content) in cell membranes (Malecka et al. 2001). In order to combating with excess ROS and protecting cells, plants synthesize a variety of non-enzymatic, low molecular weight antioxidants like, ascorbic acid (AsA), tocopherols, carotenoids, reduced glutathione (GSH), and enzymatic antioxidants like, ascorbate peroxidase (APX), peroxidase (POD), superoxide dismutase (SOD) catalase (CAT) (Gratao et al. 2005; Mishra et al. 2009).

Plants play a substantial role in remediating environmental pollution. Atmospheric Pb deposited on the surface of the soil and plants can accumulate Pb from soil. After adsorption to root surfaces, Pb enters the root passively following ionic channels/transporters (Kushwaha et al. 2018). However, plants compartmentalize the metal in the vacuole, enhancing active efflux, complexes (organic or inorganic) and induction of metal chelating stress proteins like phytochelatins, metallothioneins (Gupta et al. 2013a). In this chapter, comprehensive up to date studies are represented covering Pb uptake, transport, accumulation and Pb induced toxicity in higher plants along with a glimpse on phytoremediation strategies of plants for Pb decontamination from the environment.

2 Source of Lead

Naturally, in earth crust, lead content varies in different rocks. While, rocks like granite, black shale, and rhyolite contains approximately 30 mg kg^{-1} concentration of Pb in basalt, evaporate sediments, and the ultramafic igneous rocks (like dunite) is about 1 mg kg^{-1} (Kushwaha et al. 2018). Pb concentrations are relatively high in metamorphic and igneous rocks due to presence of high amounts of potassium (K) and silica (Si) (Wedepohl 1956). Reports suggest that some minerals of granite rocks like monazite, uraninite, xenotime, thorite, allanite, zircon, titanite those are rich in radioactive elements contain greater concentrations of Pb (Lovering 1969; Kushwaha et al. 2018).

In soil trace amounts of lead are found due to natural pedogenetic processes by weathering of parent materials (<1000 mg kg^{-1}) (Kabata-Pendias and Pendias 2001). Over the past years several anthropogenic processes are responsible for increased lead content (more than 1000-fold) in the environment. Major anthropogenic sources of Pb contamination include various daily life products based on Pb like paints (Pb based paints comprise lead chromate), Pb-based solder, Pb-glazed ceramics and Pb-containing pesticides (e.g. lead arsenate, widely used in horticulture and agriculture). Pb mining, smelting and tailings result high levels of lead contamination in soil leading to higher levels of Pb in plants grown in adjacent areas. Pb is a common contaminant in urban areas due to extensive industrialization. Vehicular emission after combustion of leaded petrol (containing tetraethyl lead) contributes substantially of Pb in urban environment. As per USEPA, permissible level of Pb in soil is 400 mg L^{-1} and acceptable limit of Pb is 0.01 mg L^{-1} in drinking water (Bureau of Indian Standard).

3 Speciation, Availability, Uptake, Transportation and Accumulation of Pb

In soil, Pb may present in a range of chemical entities having different levels of toxicity, mobility and bioavailability (Shiowatana et al. 2001) after being adsorbed by soil particles (Fig. 1). Pb distribution and migration in soil depends upon a number of reactions including mineral dissolution-precipitation, desorption-adsorption, cation exchange, organic immobilization, aqueous complexation, and uptake by available plants (Wuana and Okieimen 2011). In soil, Pb may occur as a free metal ion, in complexion with components like HCO$_3^-$, CO$_3^{2-}$, SO$_4^{2-}$, and Cl$^-$, or, may remain as organic ligands such as humic acids, fulvic acids, and amino acids. Additionally, adsorption on particle surfaces (like clay particles, organic matters or iron oxides) is common phenomenon for Pb (Uzu et al. 2009; Sammut et al. 2010). Due to strong binding affinity of Pb with organic and colloidal materials, a minor amount of Pb may be present in soluble form, which are available for plants (Kopittke et al. 2008a, b; Punamiya et al. 2010). Factors determining the speciation of Pb within soil include, soil pH, particle size, soil type, organic matters, cation

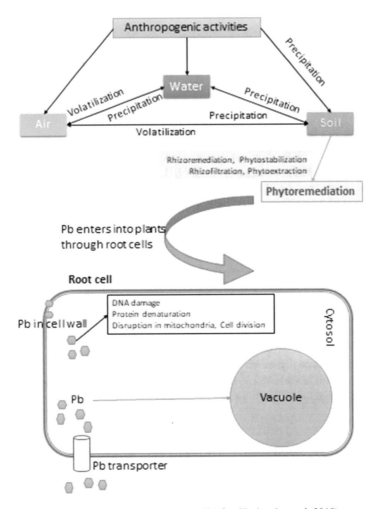

Fig. 1 Lead in environment and uptake by plant cell (after Kushwaha et al. 2018)

exchange capacity (CEC), and iron oxides (Silveira et al. 2003; Amundson et al. 2015). Similarly, availability of Pb to plants depends on both conditions of soil like pH, moisture particle size and cation exchange factors, and plant's root structure, root mycorrhiza, exudates and transpiration rate (Davies 1995). Blaylock et al. (1997) observed that major factor for Pb availability to plants is soil pH, where phosphate or carbonate precipitates control Pb solubility and availability to plants in a pH range of 5.5–7.5 and very little Pb is available to plants even if they have capability to accumulate Pb. Presence of other heavy metals in soil also have an effect on the availability of Pb. According to the report of Orrono et al. (2012), heavy metals like Zn, Cu, Ni Cd, and Cr have an antagonistic result on the Pb availability to plants. Root exudates comprising various metabolites interfere with Pb dissolution processes by modifying soil pH and by the forming of soluble organometallic compounds (Leyval and Berthelin 1991). Several organic ligands like uronates, citrates, fumarates, EDTA, and various polysaccharides cause chelation of metal ions (Mench et al. 1987). It was evidenced in plants like *Typha angustifolia* and *Vicia faba* where, considerable increase of Pb uptake by roots was found in first hour after application of EDTA (ethylenediamine tetra acetic acid) and citric acid (Muhammad et al. 2009; Shahid et al. 2012). Soil fauna residing near rhizosphere also have a profound role in metals availability and speciation as suggested from the observation in *Thlaspi caerulescens* and *Lantana camera* where accumulation of Pb within root system upsurges the occurrence of earthworm species *Pontoscolex corethrurus* (Epelde et al. 2008; Jusselme et al. 2012).

Generally, plants uptake Pb from soil and also from aerial sources. Significant quantities of Pb are taken up by plants root and bulk of it remains within root, restricting the translocation to the shoot (Lane and Martin 1977; Kumar et al. 1995). In contrary to this observation, Miller and Koeppe (1971) observed that leaf can accumulate considerable quantities of Pb in *Zea mays*. Leaf morphology (like feathery leaves) is an important factor that controls Pb absorption from aerial sources (Godzik 1993). Pb is transported into the root cell from soil through voltage gated cation channels present in plasma membrane (Huang et al. 1994). In wheat root activities of voltage gated Ca channels was found to be inhibited by Pb due to competitive inhibition of Ca^{++} for the transporter (Huang and Cunningham 1996). Similar observation was reported by Tomsig and Suszkiw (1991). The authors also reported that Pb transport through Ca^{++} channel was enhanced by a Ca^{++} channel agonist (BAY K8644) and inhibited by a Ca^{++} channel blocker (nifedipine) (Tomsig and Suszkiw 1991). In plants uptake and tolerance of Pb varies according to root system present. For example, it was observed that *Helianthus sp. and Allium sp.* with adventitious root systems (ARS) more efficiently accumulate and tolerant to Pb than the seedlings with a primary root system (PRS) (Michalak and Wierzbicka 1998; Strubińska and Hanaka 2011). The observation clearly states that adventitious root systems might have evolved mechanisms for protecting toxic effects induced by Pb.

The degree of Pb translocation from root to shoot can be described by translocation factor (TF) (the ratio between amounts of Pb in the above-ground part and in the roots) (Buscaroli 2017; Bhatti et al. 2018). Generally, the TF value for Pb is less than 1 as higher proportion (95%) of Pb gets accumulated in the roots (Chandra et al. 2018) and a small fraction (5%) gets translocated to shoots or leaves (Zhou et al. 2016; Kiran and Prasad 2017). In roots, ion-exchangeable sites present in the cell walls helps to bind Pb extracellularly. Therefore, following uptake, Pb localization is common within root due to strong binding of carboxyl group of glucuronic acid and galacturonic acid of the cell wall which restricting the apoplastic transport to the aerial parts (Rudakova et al. 1988; Inoue et al. 2013). This kind of translocation restriction phenomenon is typical to Pb, unlike other HMs (Dogan et al. 2018). Root cells in *Arabidopsis thaliana*, Pb-galacturonic acid complexes were reported following Pb treatment (Połeć-Pawlak et al. 2007). Report says that dicots accumulate higher concentration of Pb in root than the monocot (Huang and Cunningham 1996). Accumulation of Pb within root cells and lower mobility towards aerial parts also reported by several authors (Kumar et al. 2012; Gupta et al. 2013c; Dias et al. 2019) and may be an approach to avoid toxicity in the aerial parts (López-Orenes et al. 2018). However, other reasons still exist in support of lower mobility of Pb to the shoot including capturing Pb by pectines (negatively charged cell wall components), plasma accumulation, lead salt precipitation within intercellular spaces, and vacuolar sequestration in the root (Pourrut et al. 2011a). In root cells of *Raphanus sativus*, Pb^{2+} was found to attach by carboxyl groups of cell wall pectin s (Inoue et al. 2013). Accretion of Pb near the root cap causes inflation of apical meristem followed by thickening of cell wall and an upsurge of the size of vacuole (Rucińska-Sobkowiak et al. 2013). The free (precipitated) Pb ions get transported to different parts of the cell through different P-type pumps namely mitochondrial inner membrane (ATP) protein, drug resistance (PDR), and ATP-binding cassette (ABC) transporters (Lee et al. 2005). P-type transporters are associated with active efflux of Pb or Pb complexes from the cells which reduces Pb-induced toxicity by promoting sequestering the metal into inactive organelles (Jiang et al. 2017). Researchers have reported about Pb resistance among different plants that has been developed by discharging the Pb to the exterior of cell by leucine-rich repeat (LRR) transporters and ATM1 and PDR12 (Lee et al. 2005; Zhu et al. 2013).

4 Pb-Induced Toxicity in Plants

4.1 Effects on Cell Wall

Binding of Pb to the cell wall or membrane constituents leads to substantial variations in the plasticity of cell wall (Wierzbicka et al. 2007) that adversely affect the membrane potential by inducing lipid peroxidation (Yan et al. 2010). Malondialdehyde (MDA), a biomarker of lipid peroxidation is excessively produced due to lipid peroxidation (Mueller 2004), helps to crosslink and polymerizes the membrane components that have deleterious effects on cell elongation and/or division (Antosiewicz and Wierzbicka 1999). Disruption of cell wall plasticity and dislocated microtubular organization following Pb exposure revealed Pb as the antimitotic agent as reported by Antosiewicz and Wierzbicka (1999).

4.2 Inhibition of Enzyme Activity

Enzymes are the key target of all noxious HMs. Pb stress in plants interferes with the normal activities of a varied range of enzymes associated with diverse metabolic pathways. Pb having a concentration of 10^{-5} to 2.10^{-4} M leads to 50% inhibition of a number of enzymes; defined as inactivation constant (Seregin and Ivanov 2001) which is higher in Pb in comparison to other HMs (e.g. In Cd and in Ni 10^{-6} to

3.10^{-5} and 10^{-5} to 6.10^{-4}, respectively) (Sharma and Dubey 2005; Amari et al. 2017). Inhibition of the enzymes by Pb occurs by two mechanisms as suggested the literature. Although enzymatic inhibition induced by Pb is difficult to demonstrate but are evident for other cations having equivalent affinities of functional group of proteins. In the first mechanism, Pb is known to interact directly with the ligand group of the enzymes (e.g. –SH group), thereby preventing the enzyme activity by masking the catalytically active group. Besides interacting with the -SH group, blocking of –COOH group by Pb is also proved (Sharma and Dubey 2005). In the second pathway, Pb may affect absorption of vital minerals such as Zn, Fe, or Mg indispensable for metalloenzyme activities (Ros et al. 1992; Alcántara et al. 1994; Ouariti et al. 1997). It can substitute itself for the other divalent cation and thereby inactivate the enzyme specifically found in case of δ-aminolevulinate dehydratase (ALAD), a key enzyme in chlorophyll biosynthesis (Pourrut et al. 2013).

4.3   Effects on Germination and Growth

The prevention of sprouting and maturation of seedling is a general consequence of heavy metal exposure in plants (Radic et al. 2010). Lead has adverse effects on germination and maturation even at micromolar level (Kopittke et al. 2007) and lead (Pb)-emanated suppression of seed development has been evidenced in several plants like *Hordeum vulgare*, *Oryza sativa*, *Z. mays*, *Spartina alterniflora*, *Elsholtzia argyi*, *Pinus halepensis*, *Sedum alfredii* H. and *Pfaffia glomerata* (Tomulescu et al. 2004; Islam et al. 2007; Sengar et al. 2009; Gupta et al. 2010, 2011). Sprouting of seedlings and development in plants is strongly limited by lead exposure (Dey et al. 2007; Gichner et al. 2008; Gopal and Rizvi 2008). Germination may be hindered due to antagonistic activity of Pb with the enzyme amylase and protease as suggested by Sengar et al. (2009). Another study made by Yang et al. (2010) suggested that lead induced stress resulted in boosting of NADH-dependent synthesis of extracellular H_2O_2 in the growing seeds that may resulted in the arrest of wheat seed growth.

Lead restrains the growth of both underground and above-ground parts of plants at low level (Islam et al. 2007; Kopittke et al. 2007; Gupta et al. 2010), but root is more affected as it accumulates higher amounts of lead (Liu et al. 2008). Morphological effect of Pb toxicity exhibits distended inclined, and stumpy roots, profuse secondary roots per unit root length and cessation of root elongation (Kopittke et al. 2007; Arias et al. 2010). Following 48–72 h of lead exposure, mitochondrial dilation, aborted cristae, vacuoles in endoplasmic reticulum and dictyosomes, fragmented cellular membrane and hyperpycnotic nuclei were observed in the *Allium sativum root cells* (Jiang and Liu 2010). Plant biomass was shown to be repressed by higher amounts of Pb exposure (Piotrowska et al. 2009; Singh et al. 2010).

4.4   Effects on Photosynthesis

Interference in photosynthesis is a notable effect of lead phytotoxicity (Singh et al. 2010; Cenkci et al. 2010). However, disrupted photosynthetic activities are off-shoots of several indirect effects induced by Pb that may include: mutilated chloroplast ultrastructure (Elzbieta and Miroslawa 2005; Islam et al. 2007), moderate activity of two crucial chlorophyll synthesizing enzymes delta-aminolevulinic acid dehydratase (ALAD) and ferredoxin NADP⁺ reductase (Gupta et al. 2009; Cenkci et al. 2010), inhibited carotenoid and plastoquinone productions (Cenkci et al. 2010), impediment in the ETS (electron transport system) (Qufei and Fashui 2009), insufficient carbon dioxide concentration due to closure of stomata (Romanowska et al. 2006; Chen et al. 2017), reduced absorption of essential metals (Mn and Fe) (Chatterjee et al. 2004; Gopal and Rizvi 2008), and increased activity of chlorophyllase in leaves (Liu et al. 2008; Jayasri and Suthindhiran 2017). Disruption of grana (Zheng et al. 2012) and chloroplast ultrastructure (Qiao et al. 2013) due to Pb and Pb induced ROS consequences to functional inactivation by impairing electron transportation and oxygen-generating centres (Meitei et al. 2014). A recent study with lead-hyperaccumulator aquatic fern *Salvinia minima* also supported lead induced decrease in photosynthesis due to premature root membrane damage following further a lead-induced stomatal closure that resulted in economized CO_2 accessibility (Leal-Alvarado et al. 2016).

4.5   Lead-Induced Oxidative Stress

One of the significant problems of heavy metals that affect cell is the higher evolution of reactive oxygen species (ROS) (Gupta et al. 2013b). Oxidative stress is a common manifestation of metal toxicity, including Pb, generated due to the production of ROS (like hydroxyl radicals (•OH), superoxide radicals ($O_2^{\bullet-}$), and hydrogen peroxide (H_2O_2) (Pourrut et al. 2008; Liu et al. 2008; Grover et al. 2010; Singh et al. 2010; Yadav 2010). But, the degree of ROS production is reliant on the type of metal, species of metal, plant families, and duration of hazards (Pourrut et al. 2011a). The accretion of ROS inside the cells leads to oxidation of a range of biomolecules within a short period, including nucleic acids, proteins, and lipids (Yadav 2010) consequences to irreparable metabolic dysfunction, developmental alteration and genetic instability in plant species (Chakravarty and Srivastava 1992) and ultimately cell death. The major targets are histidine, arginine, lysine or proline in such oxidative transformation (Stadtman and Levine 2000; Wang et al. 2008). ROS promotes DNA base oxidation, DNA protein cross-links, generation of DNA breaks and gaps (Beckman and Ames 1997).

  As mentioned earlier, Pb causes severe alteration in lipid structures of divergent cell membranes (Grover et al. 2010; Singh et al. 2010; Yan et al. 2010), consequences in the formation of abnormal cellular structures (Gupta et al. 2009) and organelles like peroxisomes, mitochondria (Liu et al. 2008; Małecka et al. 2008) or chloroplasts (Hu et al. 2007). In Z. *mays* following lead exposure changes in lipid configuration and an outflow of potassium ion from cell were observed (Małkowski et al. 2002). Pb persuades lipid peroxidation, reducing saturated fatty acids, and increasing the unsaturated fatty acids in the membranes as observed in many plant species (Singh et al. 2010).

4.6 Genotoxicity

Pb is a potent genotoxic agent as it causes DNA damage including DNA strand-breakage, micronuclei generation (Pourrut et al. 2011b; Shahid et al. 2011; Kumar et al. 2017), chromosomal anomalies and instability, microsatellites formation (Rodriguez et al. 2013) defective cytoskeletal structure (Gichner et al. 2008), depolymerization of spindle microtubule (Malea et al. 2014). Pb induced genotoxicity may be originated indirectly or directly, through ROS production; targeting RNA and DNA molecule (Kumar and Majeti 2014; Malar et al. 2014). ROS induced formation of nicked strand and fragmented DNA is the result of structural alteration in pentose sugar or nitrogen bases by the addition of -OH group to the double bonds or the deletion of C4' hydrogen atom from deoxyribose (Martínez-Macías María et al. 2012). Furthermore, any metal that likely to induce genotoxicity merely depends on the degree of exposure, duration of the exposure and the oxidation state of the metal. Lead reduces the conformity of DNA polymerases and interrupts the function of DNA and RNA synthases (Pourrut et al. 2011a). Synergistic action of Pb with UV light and alkylating agents has been observed resulted in augmented mutagenicity in plants (Hartwig et al. 1990). In addition, Pb also induces the formation of sister-chromatid exchange (Dhir et al. 1993). Yang et al. (1999) confirmed that high amounts of Pb instigate breakage of DNA strand- and the formation of adducts (8-hydroxy-deoxyguanosine) in DNA. A recent study made by Arya et al. (2013), revealed that exposure to Pb caused severe mutilation in the roots of *Alium cepa* and *Vicia faba* also accompanied by infrequent division of cells, increase the frequency of chromosomal aberrations, DNA fragmentation and micronuclei formation in both the plants.

5 Phytoremediation of Lead

In the last two decades enormous research works has been carried out on phytotechnologies for removing of metals from terrestrial or aquatic medium. Phytoremediation techniques exploit natural hyperaccumulators or transgenic plants that can uptake and accumulate toxic HMs such as Zn, Cu, Mn, Co, Pb and Ni from 100 to 1000 times in their root or to the aerial parts compare to non-accumulator plants. A number of phytoremediation processes are applicable for removing HMs but specifically phytostabilization, phytoextraction and rhizofiltration are admissible for effective phytoremediation of Pb.

Phytostabilization is a technique by which Pb is stabilized by exploiting Pb-tolerant plant species that can efficiently uptake Pb from soil and accumulates within their root or rhizospheric region; thereby, reduces the risk of leaching and entry of Pb into ground water or into the food chain. Reports recommend that the plants having high bio-concentration factor (BF) and low translocation factor (TF) are most suitable for phytostabilization (Yoon et al. 2006). An earlier evidence on field application of Goginan grass species (*Agrostos tenuis*) for phytostabilization of Pb contaminated mine waste reported by Smith and Bradshaw (1992). Seedlings of Indian mustard *(Brassica juncea)* grown in a sand Perlite mixture treated with 625 µg g^{-1} Pb were reported to decrease Pb amount in leachate (Kumar et al. 1995). Phytostabilization property of mulberry plants has been observed by Zhou et al. (2015) and most of the Pb was found to be deposited within mulberry root when the Pb concentrations in the soil were 200, 400, 800 mg kg^{-1}, respectively. Similar observations were noted by Jiang and co-workers (2019) where concentration of Pb was higher in roots of mulberry plant than in branches, leaves or shoot.

On the other hand, in phytoextraction process Pb is removed from water or soil by plant roots and translocated to and accumulated in shoots or leaves which are harvested or incinerated subsequently (Rafati et al. 2011). Plants that have widely branched root system, fast growth rate, high biomass, tolerant to higher concentration of Pb and with a high TF value (greater than 1) are potential candidates for successful phytoextraction. Plants that are used for phytoextraction process are categorized into two groups; one group includes natural hyperaccumulators which can uptake and accumulate large amounts of Pb without any phytotoxic symptom (e.g. metallophytes); *Minuartia verna* and *Agrostis tenuis* are Pb hyperaccumulators (Fernandez et al. 2016). A recent report shows the phytoextraction capacity of *Rhus chinensis* seedlings that can concentrate more than 1000 mg kg^{-1} Pb in the shoots (Shi et al. 2018). Another group of plants having higher biomass and are able to accumulate Pb due to the induction by chelates secreted from root exudates (low molecular weight organic chelates like malic acid and citric acid) and synthetic chelates like EDTA, DTPA (diethylenetriaminepentaacetic acid) and HEDTA (N-hydroxyethylenediaminetriacetic acid) (Salt et al. 1998; Xin et al. 2015; Khan et al. 2016).

Phyzofiltration is the only method among different phytoremediation processes, in which HMs are removed from aquatic medium by using plants aquatic plants (emergent, submerged or free-floating) (Rezania et al. 2016). When considering maximum uptake of Pb by phytofiltration technique, hydroponically grown candidate plants must have thick root system and then transplanted them to the Pb polluted environment where plants can absorb and accumulate Pb in their roots and shoots (Zhu et al. 1999; Verma and Suthar 2015). Different types of phytofiltrations are known depending on the type of plant parts utilized like blastofiltration (use of seedlings), rhizofiltration (use of plant roots) and caulofiltration (use of excised plant shoots) (Mesjasz-Przybyłowicz et al. 2004). Factors including physicochemical characteristics of the plants root or photosynthetic surface microorganisms play key role on the strength of phytofiltration (Olguín and Sánchez-Galván 2012). Following uptake secreted root exudates and rhizospheric pH promote precipitation of Pb onto the root surface and saturated root surface with Pb precipitates are collected and destroyed (Flathman and Lanza 1998; Zhu et al. 1999).

6 Conclusion

In recent dates Pb is focused as a potent HMs due to huge application in the industries. Contamination of soil by Pb leads to excessive loss in crop productivity as well as alarming for human and animal health due to its release into the food chain. Pb is taken up by plant's root and remains deposited within cell walls or vacuoles as complexation with phytochelatin or glutathione. Small particles of metals also found to be sequestered within dictyosome, dictyosome derived vesicles and endoplasmic reticulum (Sharma and Dubey 2005). However, root to shoot translocation of this metal is limited due to presence of root endodermis barrier which is released at lethal Pb exposure and bring about entry of Pb into vascular tissues. Within plant cell toxic insult endorsed by Pb includes disorganization of membrane structure and loss of permeability, inhibition of enzyme activities by interaction with the thiol groups at active site, fluctuation in hormonal levels and imbalance of mineral and water homeostasis, distorted chloroplast structure, declining photosynthetic rate, alteration of oxygen evolution, obstruction in electron transport. Phenotypic manifestation of Pb exposure includes chlorosis, stunted growth and blackening of the root system.

Plants have developed different strategies to combat HMs exposure specifically non-enzymatic or enzymatic and antioxidant defence system. Key enzymes are catalase (CAT), superoxide dismutase (SOD), peroxidase (POD), ascorbate peroxidase (APX), glutathione reductase (GR) and glutathione peroxidases (GPX); directly involved in the reduction and manipulation of reactive oxygen species to maintain the cellular integrity. However, considerable investigations are needed on molecular aspects of segregation and detoxification mechanism of Pb, molecular profile of the signalling pathways and proteomic profiling of metal tolerant plants during Pb stress. Such information will support researchers to develop Pb-tolerant crops for sustainable agriculture and biological tools for remediation of contaminated lands polluted with toxic Pb.

Acknowledgements S.C. sincerely acknowledges and thanks Director, DRL (DRDO), Assam, India.

References

Alcántara E, Romera FJ, Cañete M, De la Guardia MD (1994) Effects of heavy metals on both induction and function of root Fe (lll) reductase in Fe-deficient cucumber (*Cucumis sativus* L.) plants. J Exp Bot 45:1893–1898
Amari T, Ghnaya T, Abdelly C (2017) Nickel, cadmium and lead phytotoxicity and potential of halophytic plants in heavy metal extraction. S Afr J Bot 111:99–110

Amundson R, Berhe AA, Hopmans JW, Olson C, Sztein AE, Sparks DL (2015) Soil and human security in the 21st century. Science 348:1261071

Anjum SA, Ashraf U, Khan I, Saleem MF, Wang LC (2016) Chromium toxicity induced alterations in growth, photosynthesis, gas exchange attributes and yield formation in maize. Pak J Agricult Sci 1:53

Antosiewicz D, Wierzbicka M (1999) Localization of lead in *Allium cepa* L. cells by electron microscopy. J Microsc 195:139–146

Arias JA, Peralta-Videa JR, Ellzey JT, Ren M, Viveros MN, Gardea-Torresdey JL (2010) Effects of *Glomus deserticola* inoculation on *Prosopis*: enhancing chromium and lead uptake and translocation as confirmed by X-ray mapping, ICP-OES and TEM techniques. Environ Exp Bot 68:139–148

Arshad M, Silvestre J, Pinelli E, Kallerhoff J, Kaemmerer M, Tarigo A, Shahid M, Guiresse M, Pradère P, Dumat C (2008) A field study of lead phytoextraction by various scented Pelargonium cultivars. Chemosphere 71:2187–2192

Arya SK, Basu A, Mukherjee A (2013) Lead induced genotoxicity and cytotoxicity in root cells of *Allium cepa* and *Vicia faba*. Nucleus 56:183–189

Ashraf U, Kanu AS, Deng Q, Mo Z, Pan S, Tian H, Tang X (2017) Lead (Pb) toxicity; physio-biochemical mechanisms, grain yield, quality, and Pb distribution proportions in scented rice. Front Plant Sci 8:259

Beckman KB, Ames BN (1997) Oxidative decay of DNA. J Biol Chem 32:19,633–19,636

Bhatti SS, Kumar V, Sambyal V, Singh J, Nagpal AK (2018) Comparative analysis of tissue compartmentalized heavy metal uptake by common forage crop: a field experiment. Catena 160:185–193

Blaylock MJ, Salt DE, Dushenkov S, Zakharova O, Gussman C, Kapulnik Y, Ensley BD, Raskin I (1997) Enhanced accumulation of Pb in Indian mustard by soil-applied chelating agents. Environ Sci Technol 31:860–865

Buscaroli A (2017) An overview of indexes to evaluate terrestrial plants for phytoremediation purposes. Ecol Indic 82:367–380

Cenkci S, Cigerci IH, Yildiz M, Özay C, Bozdag A, Terzi H (2010) Lead contamination reduces chlorophyll biosynthesis and genomic template stability in *Brassica rapa* L. Environ Exp Bot 67:467–473

Chakravarty B, Srivastava S (1992) Toxicity of some heavy metals in vivo and in vitro in *Helianthus annuus*. Mutat Res 283:287–294

Chandra R, Kumar V, Tripathi S, Sharma P (2018) Heavy metal phytoextraction potential of native weeds and grasses from endocrine-disrupting chemicals rich complex distillery sludge and their histological observations during in-situ phytoremediation. Ecol Eng 111:143–156

Chatterjee C, Dube BK, Sinha P, Srivastava P (2004) Detrimental effects of lead phytotoxicity on growth, yield, and metabolism of rice. Commun Soil Sci Plant Anal 35:255–265

Chen Q, Zhang X, Liu Y, Wei J, Shen W, Shen Z, Cui J (2017) Hemin-mediated alleviation of zinc, lead and chromium toxicity is associated with elevated photosynthesis, antioxidative capacity; suppressed metal uptake and oxidative stress in rice seedlings. Plant Growth Regul 81:253–264

Davies BE (1995) Lead and other heavy metals in urban areas and consequences for the health of their inhabitants. In: Majumdar SK, Miller EW, Brenner FJ (eds) Environmental contaminants, ecosystems and human health. The Pennsylvania Academy of Science, Easton, PA, pp 287–307

Dey SK, Dey J, Patra S, Pothal D (2007) Changes in the antioxidative enzyme activities and lipid peroxidation in wheat seedlings exposed to cadmium and lead stress. Braz J Plant Physiol 19:53–60

Dhir H, Roy AK, Sharma A (1993) Relative efficiency of *Phyllantus emblica* fruit extract and ascorbic acid in modifying lead and aluminium-induced sister chromatid exchanges in mouse bone marrow. Environ Mol Mutagen 21:229–236

Dias MC, Mariz-Ponte N, Santos C (2019) Lead induces oxidative stress in *Pisum sativum* plants and changes the levels of phytohormones with antioxidant role. Plant Physiol Biochem 137:121–129

Dogan M, Karatas M, Aasim M (2018) Cadmium and lead bioaccumulation potentials of an aquatic macrophyte *Ceratophyllum demersum* L.: a laboratory study. Ecotoxicol Environ Saf 148:431–440

Elzbieta W, Miroslawa C (2005) Lead-induced histological and ultrastructural changes in the leaves of soybean (*Glycine max* (L.) Merr.). Soil Sci Plant Nutr 51:203–212

Epelde L, Hernández-Allica J, Becerril JM, Blanco F, Garbisu C (2008) Effects of chelates on plants and soil microbial community: comparison of EDTA and EDDS for lead phytoextraction. Sci Total Environ 401:21–28

Fernandez S, Poschenrieder C, Marcenò C, Gallego JR, Jiménez-Gámez D, Bueno A, Afif E (2016) Phytoremediation capability of native plant species living on Pb-Zn and Hg-As mining wastes in the Cantabrian range, north of Spain. J Geochem Explor 174:10–20

Flathman PE, Lanza GR (1998) Phytoremediation: current views on an emerging green technology. J Soil Contam 7:415–432

Gichner T, Znidar I, Száková J (2008) Evaluation of DNA damage and mutagenicity induced by lead in tobacco plants. Mutat Res 652:186–190

Godzik B (1993) Heavy metals content from zinc dumps and reference areas. Pol Bot Stud 5:113–132

Gopal R, Rizvi AH (2008) Excess lead alters growth, metabolism and translocation of certain nutrients in radish. Chemosphere 70:1539–1544

Gratao PL, Polle A, Lea PJ, Azevedo RA (2005) Making the life of heavy metal-stressed plants a little easier. Funct Plant Biol 32:481–494

Grover P, Rekhadevi P, Danadevi K, Vuyyuri S, Mahboob M, Rahman M (2010) Genotoxicity evaluation in workers occupationally exposed to lead. Int J Hyg Environ Health 213:99–106

Gupta DK, Li LL (2013) Lead detoxification system in plants. In: Kretsinger RH, Uversky VN, Permyakov EA (eds) Encyclopedia of metalloproteins. Springer, Heidelberg, pp 1173–1179

Gupta DK, Nicoloso F, Schetinger M, Rossato L, Pereira L, Castro G, Srivastava S, Tripathi RD (2009) Antioxidant defense mechanism in hydroponically grown *Zea mays* seedlings under moderate lead stress. J Hazard Mater 172:479–484

Gupta DK, Huang HG, Yang XE, Razafindrabe BHN, Inouhe M (2010) The detoxification of lead in *Sedum alfredii* H. is not related with phytochelatins but the glutathione. J Hazard Mater 177:437–444

Gupta DK, Nicoloso FT, Schetinger MRC, Rossato LV, Huang HG, Srivastava S, Yang XE (2011) Lead induced responses of *Pfaffia glomerata*, an economically important Brazilian medicinal plant, under *in vitro* culture conditions. Bull Environ Contam Toxicol 86:272–277

Gupta DK, Huang HG, Corpas FJ (2013a) Lead tolerance in plants: strategies for phytoremediation. Environ Sci Pollut Res 20:2150–2161

Gupta DK, Corpas FJ, Palma JM (2013b) Heavy metal stress in plants. Springer, Heidelberg

Gupta DK, Huang HG, Nicoloso FT, Schetinger MRC, Farias JG, Li TQ, Razafindrabe BHN, Aryal N, Inouhe M (2013c) Effect of Hg, As and Pb on biomass production, photosynthetic rate, nutrients uptake and phytochelatin induction in *Pfaffia glomerata*. Ecotoxicology 22:1403–1412

Hartwig A, Schlepegrell R, Beyersmann D (1990) Indirect mechanism of lead induced genotoxicity in cultured mammalian cells. Mutat Res 241:75–82

Hu J, Shi G, Xu Q, Wang X, Yuan Q, Du K (2007) Effects of Pb^{2+} on the active oxygenscavenging enzyme activities and ultrastructure in *Potamogeton crispus* leaves. Russ J Plant Physl 54:414–419

Huang JW, Cunningham SD (1996) Lead phytoextraction: species variation in lead uptake and translocation. New Phytol 134:75–84

Huang JW, Grunes DL, Kochian LV (1994) Voltage dependent Ca^{2+} influx into right-side-out plasmamembrane vesicles isolated from wheat roots: characteristic of a putative Ca^{2+} channel. Proc Natl Acad Sci U S A 91:3473–3477

Inoue H, Fukuoka D, Tatai Y, Kamachi H, Hayatsu M, Ono M, Suzuki S (2013) Properties of lead deposits in cell walls of radish (*Raphanus sativus*) roots. J Plant Res 126:51–61

Islam E, Yang X, Li T, Liu D, Jin X, Meng F (2007) Effect of Pb toxicity on root morphology, physiology and ultrastructure in the two ecotypes of *Elsholtzia argyi*. J Hazard Mater 147:806–816

Jayasri MA, Suthindhiran K (2017) Effect of zinc and lead on the physiological and biochemical properties of aquatic plant *Lemna minor*: its potential role in phytoremediation. Appl Water Sci 7:1247–1253

Jiang W, Liu D (2010) Pb-induced cellular defense system in the root meristematic cells of *Allium sativum* L. BMC Plant Biol 10:40

Jiang L, Wang W, Chen Z, Gao Q, Xu Q, Cao H (2017) A role for APX1 gene in lead tolerance in *Arabidopsis thaliana*. Plant Sci 256:94–102

Jiang Y, Jiang S, Li Z, Yan X, Qin Z, Huang R (2019) Field scale remediation of Cd and Pb contaminated paddy soil using three mulberry (*Morus alba* L.) cultivars. Ecol Eng 129:38–44

Jusselme MD, Poly F, Miambi E, Mora P, Blouin M, Pando A, Rouland-Lefèvre C (2012) Effect of earthworms on plant *Lantana camara* Pb-uptake and on bacterial communities in root-adhering soil. Sci Total Environ 416:200–207

Kabata-Pendias A, Pendias H (2001) Trace elements in soils and plants, 3rd edn. CRC Press, Boca Raton, FL

Khan I, Iqbal M, Ashraf MY, Ashraf MA, Ali S (2016) Organic chelants-mediated enhanced lead (Pb) uptake and accumulation is associated with higher activity of enzymatic antioxidants in spinach (*Spinacea oleracea* L.). J Hazard Mater 317:352–361

Kiran BR, Prasad MNV (2017) Responses of *Ricinus communis* L. (castor bean, phytoremediation crop) seedlings to lead (Pb) toxicity in hydroponics. Selcuk J Agri Food Sci 31:73–80

Kopittke PM, Asher CJ, Kopittke RA, Menzies NW (2007) Toxic effects of Pb^{2+} on growth of cowpea (*Vigna unguiculata*). Environ Pollut 150:280–287

Kopittke PM, Asher CJ, Blamey FP, Auchterlonie GJ, Guo YN, Menzies NW (2008a) Localization and chemical speciation of Pb in roots of signal grass (*Brachiaria decumbens*) and Rhodes grass (*Chloris gayana*). Environ Sci Technol 42:4595–4599

Kopittke PM, Asher CJ, Menzies NW (2008b) Prediction of Pb speciation in concentrated and dilute nutrient solutions. Environ Pollut 153:548–554

Kumar A, Majeti NVP (2014) Proteomic responses to lead-induced oxidative stress in *Talinum triangulare* Jacq. (Willd.) roots: identification of key biomarkers related to glutathione metabolisms. Environ Sci Pollut Res 21:8750–8764

Kumar PN, Dushenkov V, Motto H, Raskin I (1995) Phytoextraction: the use of plants to remove heavy metals from soils. Environ Sci Technol 29:1232–1238

Kumar A, Prasad MN, Sytar O (2012) Lead toxicity, defense strategies and associated indicative biomarkers in *Talinum triangul* are grown hydroponically. Chemosphere 89:1056–1065

Kumar A, Pal L, Agrawal V (2017) Glutathione and citric acid modulates lead- and arsenic-induced phytotoxicity and genotoxicity responses in two cultivars of *Solanum lycopersicum* L. Acta Physiol Planta 39:151

Kushwaha A, Hans N, Kumar S, Rani R (2018) A critical review on speciation, mobilization and toxicity of lead in soil-microbe-plant system and bioremediation strategies. Ecotox Environ Safet 147:1035–1045

Lane SD, Martin ES (1977) A histochemical investigation of lead uptake in *Raphanus sativus*. New Phytol 79:281–286

Leal-Alvarado DA, Espadas-Gil F, Sáenz-Carbonell L, Talavera-May C, Santamaría JM (2016) Lead accumulation reduces photosynthesis in the lead hyper-accumulator *Salvinia minima* Baker by affecting the cell membrane and inducing stomatal closure. Aquat Toxicol 171:37–47

Lee M, Lee K, Lee J, Noh EW, Lee Y (2005) AtPDR12 contributes to lead resistance in *Arabidopsis*. Plant Physiol 138:827–836

Leyval C, Berthelin J (1991) Weathering of a mica by roots and rhizospheric microorganisms of pine. Soil Sci Soc Am J 55:1009–1016

Liu D, Li T, Jin X, Yang X, Islam E, Mahmood Q (2008) Lead induced changes in the growth and antioxidant metabolism of the lead accumulating and non-accumulating ecotypes of *Sedum alfredii*. J Integr Plant Biol 50:129–140

López-Orenes A, Dias MC, Ferrer MÁ, Calderón A, Moutinho-Pereira J, Correia C, Santos C (2018) Different mechanisms of the metalliferous *Zygophyllum fabago* shoots and roots to cope with Pb toxicity. Environ Sci Pollut Res 25:1319–1330

Lovering TG (1969) The distribution of minor elements in samples of biotite from igneous rocks-basic data. US Geological Survey

Ma Y, Egodawatta P, McGree J, Liu A, Goonetilleke A (2016) Human health risk assessment of heavy metals in urban storm water. Sci Total Environ 557:764–772

Malar S, Manikandan R, Favas PJC, Vikram Sahi S, Venkatachalam P (2014) Effect of lead on phytotoxicity, growth, biochemical alterations and its role on genomic template stability in *Sesbania grandiflora*: a potential plant for phytoremediation. Ecotoxicol Environ Saf 108:249–257

Malea P, Adamakis ID, Kevrekidis T (2014) Effects of lead uptake on microtubule cytoskeleton organization and cell viability in the seagrass *Cymodocea nodosa*. Ecotoxicol Environ Saf 104:175–181

Malecka A, Jarmuszkiewicz W, Tomaszewska B (2001) Antioxidative defense to lead stress in subcellular compartments of pea root cells. Acta Bioquim Polon 48:687–698

Małecka A, Piechalak A, Morkunas I, Tomaszewska B (2008) Accumulation of lead in root cells of Pisum sativum. Acta Physiol Plant 30:629–637

Małkowski E, Kita A, Galas W, Karcz W, Kuperberg JM (2002) Lead distribution in corn seedlings (*Zea mays* L.) and its effect on growth and the concentrations of potassium and calcium. Plant Growth Regul 37:69–76

Martínez-Macías María I, Qian W, Miki D, Pontes O, Liu Y, Tang K, Liu R, Morales-Ruiz T, Ariza R, Roldán-Arjona T, Zhu JK (2012) A DNA 3′ phosphatase functions in active DNA demethylation in *Arabidopsis*. Mol Cell 45:357–370

Meitei MD, Kumar A, Prasad MN, Malec P, Waloszek A, Maleva M, Strzałka K (2014) Photosynthetic pigments and pigment-protein complexes of aquatic plants under heavy metal stress. Photosynthetic pigments: chemical structure, biological function and ecology. Russian Academy of Sciences, St. Petersburg, Nauka, Russia, pp 314–329

Mench M, Morel JL, Guckert A (1987) Metal binding properties of high molecular weight soluble exudates from maize (*Zea mays* L.) roots. Biol Fertil Soils 3:165–169

Mesjasz-Przybyłowicz JO, Nakonieczny MI, Migula PA, Augustyniak MA, Tarnawska MO, Reimold WU, Koeberl CH, Przybyłowicz WO, Głowacka EL (2004) Uptake of cadmium lead, nickel and zinc from soil and water solutions by the nickel hyperaccumu*lator Berkheya coddii*. Acta Biol Cracov Bot 46:75–85

Michalak E, Wierzbicka M (1998) Differences in lead tolerance between *Allium cepa* plants developing from seeds and bulbs. Plant and Soil 199:251–260

Miller RJ, Koeppe DE (1971) Accumulation and physiological effects of lead in corn. In: Proceedings of University of Missouri, vol 4, Columbia, pp 186–193

Mishra M, Mishra PK, Kumar U, Prakash V (2009) NaCl phytotoxicity induces oxidative stress and response of antioxidant systems in *Cicer arietinum* L. cv. Abrodhi. Bot Res Inter 2:74–82

Mueller MJ (2004) Archetype signals in plants: the phytoprostanes. Curr Opin Plant Biol 7:441–448

Muhammad D, Chen F, Zhao J, Zhang G, Wu F (2009) Comparison of EDTA-and citric acid-enhanced phytoextraction of heavy metals in artificially metal contaminated soil by *Typha angustifolia*. Int J Phytoremediation 11:558–574

Olguín EJ, Sánchez-Galván G (2012) Heavy metal removal in phytofiltration and phycoremediation: the need to differentiate between bioadsorption and bioaccumulation. New Biotechnol 30:3–8

Orrono DI, Schindler V, Lavado RS (2012) Heavy metal availability in *Pelargonium hortorum* rhizosphere: interactions, uptake and plant accumulation. J Plant Nutr 35:1374–1386

Ouariti O, Gouia H, Ghorbal MH (1997) Responses of bean and tomato plants to cadmium: growth, mineral nutrition, and nitrate reduction. Plant Physiol Biochem 35:347–354

Piotrowska A, Bajguz A, Godlewska-Zylkiewicz B, Czerpak R, Kaminska M (2009) Jasmonic acid as modulator of lead toxicity in aquatic plant *Wolffia arrhiza* (Lemnaceae). Environ Exp Bot 66:507–513

Połeć-Pawlak K, Ruzik R, Lipiec E, Ciurzyńska M, Gawrońska H (2007) Investigation of Pb(II) binding to pectin in *Arabidopsis thaliana*. J Anal Atom Spectrom 22:968–972

Pourrut B, Perchet G, Silvestre J, Cecchi M, Guiresse M, Pinelli E (2008) Potential role of NADPH-oxidase in early steps of lead-induced oxidative burst in *Vicia faba* roots. J Plant Physiol 165:571–579

Pourrut B, Shahid M, Dumat C, Winterton P, Pinelli E (2011a) Lead uptake, toxicity, and detoxification in plants. Rev Environ Contam Toxicol 213:113–136

Pourrut B, Jean S, Silvestre J, Pinelli E (2011b) Lead-induced DNA damage in *Vicia faba* root cells: potential involvement of oxidative stress. Mutat Res 726:123–128

Pourrut B, Shahid M, Douay F, Dumat C, Pinelli E (2013) Molecular mechanism involved in lead uptake, toxicity and detoxification in higher plants. In: Gupta DK, Corpas FJ, Palma JM (eds) Heavy metal stress in plants. Springer, Heidelberg, pp 121–147

Punamiya P, Datta R, Sarkar D, Barber S, Patel M, Das P (2010) Symbiotic role of Glomus mosseae in phytoextraction of lead in vetiver grass [*Chrysopogon zizanioides* (L.)]. J Hazard Mater 177:465–474

Qiao X, Shi G, Chen L, Tian X, Xu X (2013) Lead-induced oxidative damage in sterile seedlings of *Nymphoides peltatum*. Environ Sci Pollut Res 20:5047–5055

Qufei L, Fashui H (2009) Effects of Pb^{2+} on the structure and function of photosystem II of *Spirodela polyrrhiza*. Biol Trace Elem Res 129:251–260

Radic S, Babić M, Škobić D, Roje V, Pevalek-Kozlina B (2010) Ecotoxicological effects of aluminum and zinc on growth and antioxidants in *Lemna minor* L. Ecotoxcol Environ Saf 73:336–342

Rafati M, Khorasani N, Moattar F, Shirvany A, Moraghebi F, Hosseinzadeh S (2011) Phytoremediation potential of *Populus alba* and *Morus alba* for cadmium, chromium and nickel absorption from polluted soil. Int J Environ Res 5:961–970

Rezania S, Taib SM, Din MF, Dahalan FA, Kamyab H (2016) Comprehensive review on phytotechnology: heavy metals removal by diverse aquatic plants species from wastewater. J Hazard Mater 318:587–599

Rodriguez E, Azevedo R, Moreira H, Souto L, Santos C (2013) Pb^{2+} exposure induced microsatellite instability in *Pisum sativum* in a locus related with glutamine metabolism. Plant Physiol Biochem 62:19–22

Romanowska E, Wróblewska B, Drozak A, Siedlecka M (2006) High light intensity protects photosynthetic apparatus of pea plants against exposure to lead. Plant Physiol Biochem 44:387–394

Romanowska E, Wasilewska W, Fristedt R, Vener AV, Zienkiewicz M (2012) Phosphorylation of PSII proteins in maize thylakoids in the presence of Pb ions. J Plant Physiol 169:345–352

Ros R, Morales A, Segura J, Picazo I (1992) In vivo and in vitro effects of nickel and cadmium on the plasmalemma ATPase from rice (*Oryza sativa* L.) shoots and roots. Plant Sci 83:1–6

Rucińska-Sobkowiak R, Nowaczyk G, Krzesłowska M, Rabęda I, Jurga S (2013) Water status and water diffusion transport in lupine roots exposed to lead. Environ Exp Bot 87:100–109

Rudakova EV, Karakis KD, Sidorshina ET (1988) The role of plant cell walls in the uptake and accumulation of metal ions. Fiziol Biochim Kult Rast 20:3–12

Salt DE, Smith RD, Raskin I (1998) Phytoremediation. Annu Rev Plant Biol 49:643–668

Sammut ML, Noack Y, Rose J, Hazemann JL, Proux O, Depoux M, Ziebel A, Fiani E (2010) Speciation of Cd and Pb in dust emitted from sinter plant. Chemosphere 78:445–450

Sengar RS, Gautam M, Sengar RS, Sengar RS, Garg SK, Sengar K, Chaudhary R (2009) Lead stress effects on physiobiochemical activities of higher plants. Rev Environ Contam Toxicol 196:1–21

Seregin IV, Ivanov VB (2001) Physiological aspects of cadmium and lead toxic effects on higher plants. Russ J Plant Physiol 48:523–544

Shahid M, Pinelli E, Pourrut B, Silvestre J, Dumat C (2011) Lead-induced genotoxicity to *Vicia faba* L. roots in relation with metal cell uptake and initial speciation. Ecotoxicol Environ Saf 74:78–84

Shahid M, Pinelli E, Dumat C (2012) Review of Pb availability and toxicity to plants in relation with metal speciation; role of synthetic and natural organic ligands. J Hazard Mater 15:219–220

Sharma P, Dubey RS (2005) Lead toxicity in plants. Braz J Plant Physiol 17:35–52

Shi X, Wang S, Wang D, Sun H, Chen Y, Liu J, Jiang Z (2018) Woody species *Rhus chinensis* Mill. seedlings tolerance to Pb: Physiological and biochemical response. J Environ Sci 78:63–73

Shiowatana J, McLaren RG, Chanmekha N, Samphao A (2001) Fractionation of arsenic in soil by a continuous-flow sequential extraction method. J Environ Qual 30:1940–1949

Silveira MLA, Alleoni LRF, Guilherme LRG (2003) Biosolids and heavy metals in soils. Sci Agric 60:793–806

Singh R, Tripathi RD, Dwivedi S, Kumar A, Trivedi PK, Chakrabarty D (2010) Lead bioaccumulation potential of an aquatic macrophyte *Najas indica* are related to antioxidant system. Bioresour Technol 101:3025–3032

Smith RAH, Bradshaw AD (1992) Stabilization of toxic mine wastes by the use of tolerant plant populations. Trans Inst Min Metall 81:230–237

Stadtman ER, Levine RL (2000) Protein oxidation. Ann N Y Acad Sci 899:191–208

Strubińska J, Hanaka A (2011) Adventitious root system reduces lead uptake and oxidative stress in sunflower seedlings. Biol Planta 55:771

Tomsig JL, Suszkiw JB (1991) Permeation of Pb through calcium channels: fura-2 measurements of voltage- and dihydropyridine-sensitive Pb entry in isolated bovine chromaffin cells. Biochim Biophys Acta 1069:197–200

Tomulescu IM, Radoviciu EM, Merca VV, Tuduce AD (2004) Effect of copper, zinc and lead and their combinations on the germination capacity of two cereals. J Agric Sci 15:39–42

Uzu G, Sobanska S, Aliouane Y, Pradere P, Dumat C (2009) Study of lead phytoavailability for atmospheric industrial micronic and sub-micronic particles in relation with lead speciation. Environ Pollut 157:1178–1185

Verma R, Suthar S (2015) Lead and cadmium removal from water using duckweed-*Lemna gibba* L.: impact of pH and initial metal load. Alex Eng J 54:1297–1304

Wang C, Wang X, Tian Y, Yu H, Gu X, Du W, Zhou H (2008) Oxidative stress, defense response, and early biomarkers for lead-contaminated soil in *Vicia faba* seedlings. Environ Toxicol Chem 27:970–977

Wedepohl KH (1956) Untersuchungen zur Geochemie des Bleis. Geochim Cosmochim Acta 10:69–148

Wierzbicka MH, Przedpełska E, Ruzik R, Ouerdane L, Połeć-Pawlak K, Jarosz M, Szpunar J, Szakiel A (2007) Comparison of the toxicity and distribution of cadmium and lead in plant cells. Protoplasma 231:99–111

Wuana RA, Okieimen FE (2011) Heavy metals in contaminated soils: a review of sources, chemistry, risks and best available strategies for remediation. ISRN Ecol 2011:402647

Xin J, Huang B, Dai H, Zhou W, Yi Y, Peng L (2015) Roles of rhizosphere and root-derived organic acids in Cd accumulation by two hot pepper cultivars. Environ Sci Pollut Res 22:6254–6261

Yadav S (2010) Heavy metals toxicity in plants: an overview on the role of glutathione and phytochelatins in heavy metal stress tolerance of plants. South Afr J Bot 76:167–179

Yan ZZ, Ke L, Tam NFY (2010) Lead stress in seedlings of *Avicennia marina*, a common mangrove species in South China, with and without cotyledons. Aquat Bot 92:112–118

Yang JL, Wang LC, Chang CY, Liu TY (1999) Singlet oxygen is the major participating in the induction of DNA strand breakage and 8-hydroxydeoxyguanosine adduct by lead acetate. Environ Mol Mutagen 33:194–201

Yang Y, Wei X, Lu J, You J, Wang W, Shi R (2010) Lead-induced phytotoxicity mechanism involved in seed germination and seedling growth of wheat (*Triticum aestivum* L.). Ecotox Environ Safet 73:1982–1987

Yoon J, Cao X, Zhou Q, Ma LQ (2006) Accumulation of Pb, Cu, and Zn in native plants growing on a contaminated Florida site. Sci Total Environ 368:456–464

Zheng L, Peer T, Seybold V, Lütz-Meindl U (2012) Pb-induced ultrastructural alterations and subcellular localization of Pb in two species of Lespedeza by TEM-coupled electron energy loss spectroscopy. Environ Exp Bot 77:196–206

Zhou L, Zhao Y, Wang S, Han S, Liu J (2015) Lead in the soil–mulberry (*Morus alba* L.)-silkworm (*Bombyx mori*) food chain: translocation and detoxification. Chemosphere 128:171–177

Zhou C, Huang M, Li Y, Luo J, Ping Cai L (2016) Changes in subcellular distribution and anti-oxidant compounds involved in Pb accumulation and detoxification in *Neyraudia reynaudiana*. Environ Sci Pollut Res 23:21794–21804

Zhu YL, Zayed AM, Quian JH, De Souza M, Terry N (1999) Phytoaccumulation of trace elements by wetland plants: II. Water hyacinth. J Environ Qual 28:339–344

Zhu FY, Li L, Lam PY, Chen MX, Chye ML, Lo C (2013) Sorghum extracellular leucine-rich repeat protein SbLRR2 mediates lead tolerance in transgenic *Arabidopsis*. Plant Cell Physiol 54:1549–1559

Mechanisms Involved in Photosynthetic Apparatus Protection Against Lead Toxicity

Krzysztof Tokarz, Barbara Piwowarczyk, and Wojciech Makowski

Abstract Lead is one of the most toxic trace elements influencing plant growth and development processes including photosynthesis. Although lead is characterized by low mobility, it can get into the plant and after destroying physical barrier (Casparian strip), enters the xylem and translocates to the aerial parts of a plant affecting photosynthetic apparatus in both indirect and direct ways.

Directly Pb impacts on the number and ultrastructure of chloroplasts, the photosynthetic pigments synthesis, activity and efficiency of Oxygen Evolving Complex, Photosystem II Reaction Center, membrane transporters, Thioredoxin System and the Calvin-Benson Cycle effectiveness as well as synthesis and distribution of carbohydrates. Indirectly, Pb impairs plants' redox state, leading to ROS generation, wherein the photosynthetic apparatus might be source as well as objective of ROS action. In order to survive in such conditions, plants have developed a number of mechanisms protecting their photosynthetic apparatus against toxic effects of Pb.

Keywords Lead · Photosynthetic apparatus · Photosynthesis · Photosystem II · Electron transport · Chloroplast · ROS

1 Introduction

Lead (Pb) is a heavy metal widely distributed in the nature. It cannot be decomposed in the environment and it is not biodegradable, moreover Pb accumulates in the tissues of living organisms. Although, once it was widely used, now it is considered to be a big risk for human health. The development of civilization and the associated increase of Pb emissions lead to a significant contamination of the environment (Cheng and Hu 2010). On an industrial scale, Pb is used for the production

K. Tokarz (✉) · B. Piwowarczyk · W. Makowski
University of Agriculture in Krakow, Institute of Plant Biology and Biotechnology, Kraków, Poland
e-mail: krzysztof.tokarz@urk.edu.pl

© Springer Nature Switzerland AG 2020
D. K. Gupta et al. (eds.), *Lead in Plants and the Environment*, Radionuclides and Heavy Metals in the Environment,
https://doi.org/10.1007/978-3-030-21638-2_7

117

of paints, batteries, accumulators, ammunition, and so on. Although in recent times Pb contamination of the environment has been severely limited by withdrawal of this element as a gasoline additive, the threat of its toxicity is still present. According to the Report of the European Environment Agency (EEA) (EEA 2016), although the concentration of Pb in the atmosphere is low, this element is still deposited in the soil from where it can be taken up by plants and get to the food chain. In agricultural soils, Pb can originate from pesticides, inorganic fertilizers, especially phosphates (Atafar et al. 2010). According to Slootweg et al. (2010) in Europe there are about 20% of the areas threatened by lead deposition above the critical level.

2 Pb Toxicity for Plants

Pb belongs to the elements displaying the translocation restriction phenomenon (Dogan et al. 2018), which means that most of plant available Pb is retained and accumulated within the roots (Kumar and Prasad 2018). However, a small amount of Pb ions can also be transported (mostly by xylem vessels) to above-ground plant parts (Zhou et al. 2016). The first biochemical reaction to the Pb presence inside the cell is the disturbance of the redox homeostasis, although Pb as a metal has low redox activity and does not generate reactive oxygen species (ROS) directly in the Haber-Weiss/Fenton process (Shahid et al. 2014, 2015). Excessive ROS production due to Pb toxicity is primarily result of enzyme damage either by replacing divalent cations at the cofactor sites or by altering enzyme activities by attaching to their –SH groups (Shahid et al. 2014, 2015). In excess, ROS can affect polyunsaturated fatty acids and cause lipid peroxidation and damage of the cell membranes (Hattab et al. 2016; Ashraf and Tang 2017). Moreover, Pb can directly or indirectly, through ROS generation, also damage RNA, DNA or cause protein oxidation (Kumar and Prasad 2018). In addition, in the green parts of the plant Pb can have direct or indirect major impact on the photosynthetic process (Chen et al. 2017; Kumar and Prasad 2018).

3 Effects of Lead Toxicity on Photosynthetic Apparatus and Photosynthetic Process

3.1 Pigments Biosynthesis

One of the visible signs of toxic effects of Pb ions on the photosynthetic apparatus is chlorosis (Ghori et al. 2019) caused by the decrease of chlorophyll content in plants what was observed in *Anthyllis vulneraria* (Piwowarczyk et al. 2018), *Rhus chinensis* (Shi et al. 2019), *Helianthus annuus* (Doncheva et al. 2018). The toxicity of Pb is associated with the inactivation of 5-aminolevulinic acid dehydratase

(ALAD)—enzyme which is necessary for the synthesis of porphobilinogen—the precursor of chlorophyll. This inactivation is achieved through Pb interaction with the –SH groups of the enzyme active side (Prasad and Prasad 1987; Sharma and Dubey 2005). In addition, Pb ions substitute Mg ions in the porphyrin ring of chlorophyll molecule leading to impaired function (Harpaz-Saad et al. 2007). Often, the decrease in the content of chlorophyll pigments is accompanied by a decrease in the content of carotenoid pigments (Ashraf and Tang 2017; Kaviani et al. 2017). These pigments have a dual function in the process of photosynthesis: firstly—they participate in the mechanism of harvesting additional energy not absorbed by chlorophyll and, due to their structure. Secondly—they protect against the negative effect of excess energy reaching the reaction center in the process of dissipating absorbed energy as a heat (Sun et al. 2018). The toxic action of Pb ions mainly arises from the inactivation of enzymes associated with carotenoid synthesis pathway (MEP-methyl eritrythol phosphate pathway) that takes place in the chloroplasts (Giuliano 2014).

3.2 Chloroplast Ultrastructure

The effect of Pb on the chloroplast ultrastructure depends to a greater or lesser extent on the concentration used in the experiment. At low applied Pb concentrations, chloroplasts morphologically do not differ from control plants chloroplasts or show some minor ultrastructural changes (Arena et al. 2017; Figlioli et al. 2019). Higher Pb concentrations lead to chloroplast shrinkage (Khan et al. 2018) or their swelling (Shen et al. 2016; Zhou et al. 2017). In addition, Pb ions can also lead to a less developed thylakoid system manifested by a decrease in the amount of granal thylakoids and an increase in the amount of stromal thylakoids, as well as a decrease in the compaction level of thylakoids in the granum, and thus their loosening (Reis et al. 2015; Shen et al. 2016; Zhou et al. 2017; Khan et al. 2018). Furthermore, due to Pb toxicity some authors observe the lack of starch granules (Zhou et al. 2017) or the increase in their number (Khan et al. 2018; Figlioli et al. 2019). Moreover, in chloroplasts exposed to Pb ions, an increase in the number and size of osmiophilic granules (Shen et al. 2016) and among them plastoglobules (Piwowarczyk et al. 2018) can be observed. The toxicity of Pb to chloroplasts is associated with the disruption in their ultrastructure, which results from the high affinity of Pb ions for the amino and thiol groups of structural thylakoid membrane proteins (Xiong et al. 2006; Hu et al. 2007; Liu et al. 2008; Piotrowska et al. 2009). Moreover, these structural changes of thylakoid membrane proteins can lead to a decrease or complete loss of chloroplast membranes selective permeability. Loss of selective permeability leads to an uncontrolled flow of ions across the membranes, making it impossible to generate the proton gradient necessary for the ATP synthesis.

3.3   Light and Dark Reactions of Photosynthesis

3.3.1   Light Reactions

In the light reactions, the toxic effects of Pb are mainly related to the activity of the oxygen evolving complex (OEC), the reaction center of photosystem II (PSII), the electron transport chain, including membrane transporters, photosystem I (PSI), Ferredoxine-NADP + reductase and ATP synthase (Fig. 1) (Gupta et al. 2009; Cenkci et al. 2010; Kalaji et al. 2016).

Sersen et al. (2014) observed disorders at the donor side of PSII in isolated spinach chloroplasts in the presence of Pb caused by irreversible damage to the OEC. The direct toxicity of Pb is associated with Zn^{2+} and/or Ca^{2+} substitution by Pb^{2+} cations. In addition, Pb ions may associate with the thiol groups of the complex, leading to disturbances in its activity as a result of conformational changes (Rashid et al. 1994; Sersen et al. 2014). Next target of Pb toxicity in light reactions is light harvesting complex (LHC). Janik et al. (2013) have pointed out, that decreased electron

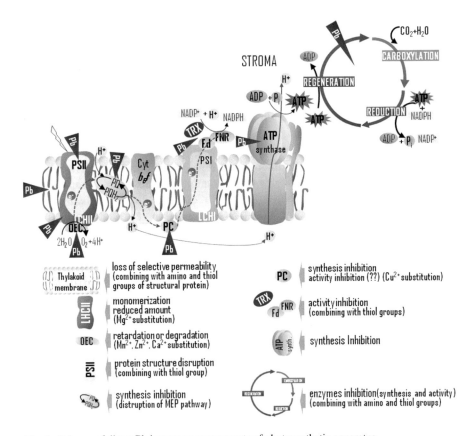

Fig. 1 Scheme of direct Pb impact on components of photosynthetic apparatus

transport efficiency may be a consequence of monomerization of LHCII protein, caused by the presence of Pb ions.

In the reaction centers of PSII, the direct Pb effect is related to the substitution of Mg ions by Pb ions in the porphyrin ring of chlorophyll *a* that is a cortical part of the PSII reaction centre (Harpaz-Saad et al. 2007; Bechaieb et al. 2016). The chlorophyll molecule formed in this way is in a state of constant excitation leading to permanent inactivation of the PSII reaction centre (Sharma and Dubey 2005; Romanowska et al. 2008). In addition, the association of Pb ions with the Q_B site on the D1 protein is observed, as well as their association with the -SH groups of the PSII complex proteins. This leads to changes in the secondary and tertiary structure of proteins, the effect of which is the inhibition of electron transport outside PSII (Qufei and Fashui 2009). Another place of Pb impact is the acceptor side of PSII. Pb limits the synthesis of plastoquinone by disturbing the enzymes involved in the pathway of its synthesis (MEP-IPP isopentenyl diphosphate pathway) (Schultz et al. 1985; Giuliano 2014). The direct effect of Pb on plastocyanin is limitation of its synthesis and changing its activity due to substitution of copper ions by Pb ions. On the other hand, Pb directly interacts with NADP+ ferredoxin reductase by combining with its thiol groups, which leads to the limitation of the linear electron transport rate and suppression of the reduction power (NADPH) synthesis (Doncheva et al. 2018).

At the same time, the indirect impact on the electron chain is associated with the generation of ROS by the PSII as a result of: limiting of the OEC activity and thus limiting of the amount of electrons supplied to PSII reaction centres. Lack of photochemical use of excited electrons from the reaction centre can generate ROO· radicals while oxidized P680 also generates ROS (Tyystjärvi 2008). The indirect influence of Pb on PS I is manifested by PSI over reduction caused by limitation of activity on the acceptor side of this photosystem (thioredoxins, especially the ferredoxin-NADP + reductase) (Stefanov et al. 2014; Doncheva et al. 2018).

3.3.2   Dark Reactions

Pb can also cause inhibition of photosynthesis through reducing activity of Calvin cycle enzymes (Liu et al. 2008; Singh et al. 2010). Hampp et al. (1973) reported, that in presence of Pb enzymes involved in photosynthetic CO_2 assimilation: ribulose-1,5-bisphosphate carboxylase (Rubisco) and ribulose-1,5-bisphosphate kinase were inhibited. Pb ions can affect activity, but also the quantity of enzymatic proteins. Arena et al. (2017) have shown that plants of *Cynara cardunculus* L. growing under Pb stress accumulated less Rubisco, than control plants.

3.4   Thioredoxin System

A number of mechanisms regulate photochemical use efficiency of absorbed radiation in the photosynthesis process and at the same time protect against the appearance of ROS. These include: nonphotochemical quenching (NPQ), electrons flow control system between PSII and PSI, state transitions, cycling and pseudocycling electron flow pathways, and a system of activation and deactivation of enzymes involved in both light and carbon fixation reactions. In addition, in the plant cells there is a system of thioredoxin proteins responsible for a number of mechanisms regulating photosynthesis in chloroplasts (Nikkanen et al. 2017). Thioredoxins, a group of proteins with oxidoreductive properties, mediate the reduction of other proteins by its oxidation. They are reduced by thioredoxin reductases with which they form a common thioredoxin system. Most of the thioredoxins are located in chloroplasts, and electron donors are ferredoxin-dependent, Fd-thioredoxin reductase (FTR) and NADPH-dependent, NADP-thioredoxin reductase complex (NTRC) (Balsera et al. 2014; Geigenberger and Fernie 2014; Nikkanen and Rintamäki 2014; Rouhier et al. 2015; Buchanan 2016; Nikkanen et al. 2017). In photosynthesizing cells, FTR-dependent system is activated with light at the time of ferredoxin reduction by PS I, while the NTRC dependent system is activated both by light (NADPH from the light phase) and darkness (NADPH from the pentose-phosphate cycle) (Motohashi and Hisabori 2010; Geigenberger and Fernie 2014; Nikkanen and Rintamäki 2014; Bölter et al. 2015; Rouhier et al. 2015; Kang and Wang 2016; Nikkanen et al. 2017). The FTR-dependent system is directly related to plant productivity by activating enzymes in the Calvin-Benson cycle (Buchanan 2016; Nikkanen et al. 2017). In contrast, the NTRC-dependent system participates in the control of chloroplast development in response to changes in spectral composition and intensity of radiation, signal transfer between chloroplast compartments, as well in the antioxidant system (Motohashi and Hisabori 2010; Geigenberger and Fernie 2014; Nikkanen and Rintamäki 2014; Bölter et al. 2015; Rouhier et al. 2015; Kang and Wang 2016; Nikkanen et al. 2017). Molecular oxygen derived from the OEC activity, oxidizes the enzymes involved in the photosynthetic process, hence the high activity of thioredoxins is necessary to maintain these enzymes' activity ensuring the process of photosynthesis. Pb, that has a high affinity to thiol groups, combining with the elements of the thioredoxin system, diminishes the reduction potential necessary to maintain the high efficiency of the photosynthesis process. In addition, thioredoxin f (TRXf), dependent on FTR and NTRC, is involved in the control of ATP synthase as well as redox activated enzymes controlling chlorophyll synthesis, shikimate pathway, Calvin-Benson cycle and starch synthesis (Geigenberger and Fernie 2014; Nikkanen and Rintamäki 2014; Nikkanen et al. 2017). Thus, the inactivation of FTR, NTRC and TRXf by Pb ions will result in reduced synthesis of chlorophylls and ATP, as well as in decreased efficiency of the dark photosynthetic phase.

4   Plant Response to Lead Toxicity

Plants have developed several strategies that allow them to avoid the stress associated with the appearance of heavy metals (including Pb) in the environment. These strategies can be divided into mechanism of stress avoidance and stress tolerance. The avoidance strategy consisting in preventing the entry of Pb into protoplasts of metabolically active cells is based on the mechanism of exclusion and accumulation (compartmentalization) (Krzesłowska et al. 2016; Dresler et al. 2017). In the exclusion mechanism, Pb is immobilized in the rhizosphere by root mucilage rich in organic acids, pectins and other compounds that bind metal ions (Sobotik et al. 1998) as well as captured by border cells present in root exudates (Huskey et al. 2018). In the accumulation mechanism, Pb is retained in plant parts with low biological activity mainly the cell walls (Krzesłowska et al. 2016). In turn, if Pb enters the cells, plant sets tolerance/defence mechanisms based on interlinked molecular and physiological processes leading to effective cells detoxification and/or repairing damages caused by Pb ions (Krzesłowska et al. 2016; Dresler et al. 2017). Detoxification is based on the binding of Pb with various ligands: phytochelatins, organic acids, amino acids, metallothioneins, phenols and their transport into various parts of the cell, especially to vacuoles (Krzesłowska et al. 2016). In addition, oxidative stress resulting from the presence of Pb ions stimulates the activity of the plant antioxidant system including enzymes (superoxide dismutase, ascorbate peroxidase, glutathione peroxidase, glutathione reductase, catalase) and small molecules (glutathione, cysteines, phenolic compounds) that scavenge ROS (Dresler et al. 2017; Shahid et al. 2017; Zhong et al. 2017).

4.1 Mechanisms of Photosynthetic Apparatus Protection Against Lead Toxicity

PSII is the plant photosynthetic apparatus element highly exposed to the Pb toxic effects. The presence of Pb affects directly and indirectly its functioning, leading to the emergence of ROS. The mechanism protecting the photosynthetic apparatus from the toxic effects of Pb is associated with the limitation of the amount of radiation absorbed, its effective dissipation by the PSII reaction center and more efficient transport of the electron outside the PSII reaction center (Fig. 2). Research of Piwowarczyk et al. (2018) showed that *Anthyllis vulneraria* treated with Pb was characterized by a reduced chlorophyll content and chl *a/b* ratio pointing the diminish size of photosynthetic antennas and the number of active PSII reaction. These changes protect against excessive radiation, allowing the absorption spectrum of antennas to be balanced. This excludes the risk of disturbances in the amount of absorbed radiation resulting from changes in the spectral composition of absorbed radiation (Ruban and Johnson 2009). Also, Arena et al. (2017) demonstrated that Pb presence in growing medium lead to decreased quantum yield performance, and increase in non-photochemical quenching in compare to control plants, with no differences in PSII maximal photochemical efficiency. Such results may indicate that in plants growing under Pb presence, rise in non-photochemical quenching is effective mechanism of protecting the photosystems from excess energy.

Lead ions have low translocation ability from roots to shoots, but regardless of that they can change phosphorylation level of proteins involved in photosystems construction (Arena et al. 2017). In the study of Wasilewska et al. (2015), pea plants exposed to

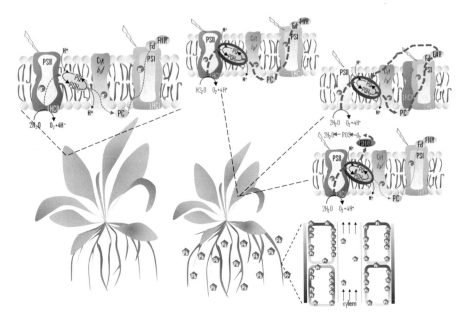

Fig. 2 Scheme of potential mechanisms of photosynthetic apparatus protection against lead toxicity (Adopted from Doncheva et al. 2018; Piwowarczyk et al. 2018)

Pb were characterized by increased phosphorylation of D1 protein, with simultaneously lover phosphorylation level of LHC II, in compare to the control plants. Arena et al. (2017) have shown that lead stress increase accumulation of D1 protein in *C. cardunculus*. More, Romanowska et al. (2012) have reported that heavy metal stress increased phosphorylation of the PSII core proteins. Higher phosphorylation of D1 may stabilized the structure of PS II in stress conditions, while low phosphorylation of LHC II contribute to better energy distribution between PS I and PS II (Wasilewska et al. 2015). Also, more phosphorylated D1 can be related to prevention in superoxide anion production on PS II, what decrease the oxidative stress level (Wasilewska et al. 2015). Romanowska et al. (2012) and Wasilewska et al. (2015) indicated that such changes in phosphorylation status of proteins in PS II are the universal mechanism improving membranes flexibility and photosystems stability under heavy metal stress in plants. Stability of PSII functioning in the presence of heavy metals depends on efficient and effective electron transport beyond Q_B towards linear, as well as alternative electron transport pathways. In the studies of Stefanov et al. (2014) sunflower plants characterized by high efficiency of electron transport did not show symptoms of Pb toxicity. Similar observations were noted by Doncheva et al. (2018). However, the presence of Pb in chloroplasts results in chloroplast ultrastructure changes and overreduction of acceptor side of PSI leading to impaired speed and efficiency of linear electron transport (LEF). However, altered chloroplast ultrastructure consisting of a reduction in the number of PSII complexes with a simultaneous increase in the amount of granal and stromal lamellae plastoquinone pools allows to increase the yield of both proton gradient regulation complexes (PGR5/PGRL1) and NAD(P)H dehydrogenase complex (NDH) dependent cycling electron flow (CEF). All this increases the NPQ efficiency and enables protection against PSII over reduction (Wood et al. 2018). Moreover in the studies of Doncheva et al. (2018) in the chloroplasts of sunflower plants subjected to high Pb concentration, an increase of plastid terminal oxidase (PTOX) activity and water–water cycle was observed, which compensated the reduction of linear electron flow caused by decreased ferredoxin activity and the TRX system by Pb ions (Fig. 2). In research of Piwowarczyk et al. (2018) effective protection of photosystem II against Pb was associated with enlargement of plastoquinone pool and increased intensity of cyclic electron transport, which was accompanied by LHCII size reduction. These results indicate that plant protection mechanisms against Pb toxic effects are associated with the development of the most optimal perception and utilization methods of absorbed radiation what allows for protection against ROS generation.

5 Concluding Remarks

Among many factors affecting the process of photosynthesis, the presence of Pb significantly limits the efficiency of the photosynthetic apparatus. Toxic effects of Pb are associated with the direct impact on the number and ultrastructure of chloroplasts, the photosynthetic pigments synthesis, the efficiency of light phase of photosynthesis, the Calvin-Benson cycle effectiveness as well as synthesis and distribution of carbohydrates. Indirectly, Pb disturbs the plant mineral and water metabolisms leading to the creation of an energy imbalance in the plant resulting in ROS generation. The source as well as objective of ROS action might be photosynthetic apparatus.

Plants have developed a number of mechanisms to protect the photosynthetic apparatus from the adverse effects of Pb, which reduce the risk of PSII over reduction by limiting the amount of light energy reaching the PSII reaction center, its effective non-photochemical dissipation, efficient distribution to LEF, CEF and alternative cycles as well as effective ROS scavenging by antioxidant system.

However, the effectiveness and efficiency of plant protective mechanisms in various plant species and even different varieties remains open. Knowledge and understanding the plant mechanisms that effectively counteract the toxic effects of lead on the photosynthetic apparatus would enable the use of many plant species for phytoremediation of contaminated post-industrial areas.

References

Arena C, Figlioli F, Sorrentino MC, Izzo LG, Capozzi F, Giordano S, Spagnuolo V (2017) Ultrastructural, protein and photosynthetic alterations induced by Pb and Cd in *Cynara cardunculus* L., and its potential for phytoremediation. Ecotox Environ Saf 145:83–89

Ashraf U, Tang X (2017) Yield and quality responses, plant metabolism and metal distribution pattern in aromatic rice under lead (Pb) toxicity. Chemosphere 176:141–155

Atafar Z, Mesdaghinia A, Nouri J, Homaee M, Yunesian M, Ahmadimoghaddam M, Mahvi AH (2010) Effect of fertilizer application on soil heavy metal concentration. Environ Monit Assess 160:83–89

Balsera M, Uberegui E, Schuermann P, Buchanan BB (2014) Evolutionary development of redox regulation in chloroplasts. Antioxid Redox Sign 21:1327–1355

Bechaieb R, Akacha AB, Gérard H (2016) Quantum chemistry insight into Mg-substitution in chlorophyll by toxic heavy metals: Cd, Hg and Pb. Chem Phys Lett 663:27–32

Bölter B, Soll J, Schwenkert S (2015) Redox meets protein trafficking. BBA-Bioenergetics 1847:949–956

Buchanan BB (2016) The path to thioredoxin and redox regulation in chloroplasts. Annu Rev Plant Biol 67:1–24

Cenkci S, Ciğerci İH, Yıldız M, Özay C, Bozdağ A, Terzi H (2010) Lead contamination reduces chlorophyll biosynthesis and genomic template stability in Brassica rapa L. Environ Exp Bot 67:467–473

Chen Q, Zhang X, Liu Y, Wei J, Shen W, Shen Z, Cui J (2017) Hemin-mediated alleviation of zinc, lead and chromium toxicity is associated with elevated photosynthesis, antioxidative capacity; suppressed metal uptake and oxidative stress in rice seedlings. Plant Growth Regul 81:253–264

Cheng H, Hu Y (2010) Lead (Pb) isotopic fingerprinting and its applications in lead pollution studies in China: a review. Environ Pollut 158:1134–1146

Dogan M, Karatas M, Aasim M (2018) Cadmium and lead bioaccumulation potentials of an aquatic macrophyte Ceratophyllum demersum L.: a laboratory study. Ecotox Environ Safe 148:431–440

Doncheva S, Ananieva K, Stefanov D, Vassilev A, Gesheva E, Dinev N (2018) Photosynthetic electron transport and antioxidant defense capacity of sunflower plants under combined heavy metal stress. Genet Plant Physiol 8:3–23

Dresler S, Wójciak-Kosior M, Sowa I, Stanisławski G, Bany I, Wójcik M (2017) Effect of short-term Zn/Pb or long-term multi-metal stress on physiological and morphological parameters of metallicolous and nonmetallicolous Echium vulgare L. populations. Plant Physiol Biochem 115:380–389

EEA (2016) Air quality in Europe-2016 report, Report No 28/2016, European Environment Agency, Luxembourg: Publications Office of the European Union, http://powietrze.gios.gov.pl

Figlioli F, Sorrentino MC, Memoli V, Arena C, Maisto G, Giordano S, Capozzi F, Spagnuolo V (2019) Overall plant responses to Cd and Pb metal stress in maize: growth pattern, ultrastructure, and photosynthetic activity. Environ Sci Pollut Res 26:1781–1790

Geigenberger P, Fernie AR (2014) Metabolic control of redox and redox control of metabolism in plants. Antioxid Redox Sign 21:1389–1421

Ghori NH, Ghori T, Hayat MQ, Imadi SR, Gul A, Altay V, Ozturk M (2019) Heavy metal stress and responses in plants. Int J Environ Sci Te 16:1807–1828

Giuliano G (2014) Plant carotenoids: genomics meets multi-gene engineering. Curr Opin Plant Biol 19:111–117

Gupta DK, Nicoloso FT, Schetinger MRC, Rossato LV, Pereira LB, Castro GY, Srivastava S, Tripathi RD (2009) Antioxidant defense mechanism in hydroponically grown Zea mays seedlings under moderate lead stress. J Hazard Mater 172:479–484

Hampp R, Zeigler H, Zeigler I (1973) Influence of lead ions on the activity of enzymes of reductive pentose phosphate pathway. Biochem Physiol Pflanz 164:588–495

Harpaz-Saad S, Azoulay T, Arazi T, Ben-Yaakov E, Mett A, Shiboleth YM, Hörtensteiner S, Gidoni D, Gal-On A, Goldschmidt EE, Eyal Y (2007) Chlorophyllase is a rate-limiting enzyme in chlorophyll catabolism and is posttranslationally regulated. Plant Cell 19:1007–1022

Hattab S, Hattab S, Flores-Casseres ML, Boussetta H, Doumas P, Hernandez LE, Banni M (2016) Characterisation of lead-induced stress molecular biomarkers in Medicago sativa plants. Environ Exp Bot 123:1–12

Hu JZ, Shi GX, Xu QS, Wang X, Yuan QH, Du KH (2007) Effects of Pb^{2+} on the active oxygen-scavenging enzyme activities and ultrastructure in Potamogeton crispus leaves. Russ J Plant Physiol 54:414–419

Huskey DA, Curlango-Rivera G, Root RA, Wen F, Amistadi MK, Chorover J, Hawes MC (2018) Trapping of lead (Pb) by corn and pea root border cells. Plant Soil 430:205–217

Janik E, Szczepaniuk J, Maksymiec W (2013) Organization and functionality of chlorophyll-protein complexes in thylakoid membranes isolated from Pb-treated *Secale cereale*. J Photoch Photobio B 125:98–104

Kalaji HM, Jajoo A, Oukarroum A, Brestic M, Zivcak M, Samborska IA, Cetner MD, Łukasik I, Goltsev V, Ladle RJ (2016) Chlorophyll a fluorescence as a tool to monitor physiological status of plants under abiotic stress conditions. Acta Physiol Plant 38:102

Kang ZH, Wang GX (2016) Redox regulation in the thylakoid lumen. J Plant Physiol 192:28–37

Kaviani E, Niazi A, Heydarian Z, Moghadam A, Ghasemi-Fasaei R, Abdollahzadeh T (2017) Phytoremediation of Pb-contaminated soil by *Salicornia iranica*: key physiological and molecular mechanisms involved in Pb detoxification. Clean–Soil Air Water 45:1500964

Khan MM, Islam E, Irem S, Akhtar K, Ashraf MY, Iqbal J, Liu D (2018) Pb-induced phytotoxicity in para grass (*Brachiaria mutica*) and Castorbean (*Ricinus communis* L.): antioxidant and ultrastructural studies. Chemosphere 200:257–265

Krzesłowska M, Rabęda I, Basińska A, Lewandowski M, Mellerowicz EJ, Napieralska A, Samardakiewicz S, Woźny A (2016) Pectinous cell wall thickenings formation–a common defense strategy of plants to cope with Pb. Environ Pollut 214:354–361

Kumar A, Prasad MNV (2018) Plant-lead interactions: transport, toxicity, tolerance, and detoxification mechanisms. Ecotoxicol Environ Saf 166:401–418

Liu D, Li TQ, Jin XF, Yang XE, Islam E, Mahmood Q (2008) Lead induced changes in the growth and antioxidant metabolism of the lead accumulating and non-accumulating ecotypes of *Sedum alfredii*. J Integr Plant Biol 50:129–140

Motohashi K, Hisabori T (2010) CcdA is a thylakoid membrane protein required for the transfer of reducing equivalents from stroma to thylakoid lumen in the higher plant chloroplast. Antioxid Redox Sign 13:1169–1176

Nikkanen L, Rintamäki E (2014) Thioredoxin-dependent regulatory networks in chloroplasts under fluctuating light conditions. Philos T Roy Soc B 69:20130224

Nikkanen L, Toivola J, Diaz MG, Rintamäki E (2017) Chloroplast thioredoxin systems: prospects for improving photosynthesis. Philos T Roy Soc B 372:20160474

Piotrowska A, Bajguz A, Godlewska-Żyłkiewicz B, Czerpak R, Kamińska M (2009) Jasmonic acid as modulator of lead toxicity in aquatic plant *Wolffia arrhiza* (Lemnaceae). Environ Exp Bot 66:507–513

Piwowarczyk B, Tokarz K, Muszyńska E, Makowski W, Jędrzejczyk R, Gajewski Z, Hanus-Fajerska E (2018) The acclimatization strategies of kidney vetch (*Anthyllis vulneraria* L.) to Pb toxicity. Environ Sci Pollut Res 25:19739–19752

Prasad DDK, Prasad ARK (1987) Effect of lead and mercury on chlorophyll synthesis in mung bean seedlings. Phytochemistry 26:881–883

Qufei L, Fashui H (2009) Effects of Pb^{2+} on the structure and function of photosystem II of *Spirodela polyrrhiza*. Biol Trace Elem Res 129:251

Rashid A, Camm EL, Ekramoddoullah AK (1994) Molecular mechanism of action of Pb^{2+} and Zn^{2+} on water oxidizing complex of photosystem II. FEBS Lett 350:296–298

Reis GSM, de Almeida AAF, de Almeida NM, de Castro AV, Mangabeira PAO, Pirovani CP (2015) Molecular, biochemical and ultrastructural changes induced by Pb toxicity in seedlings of *Theobroma cacao* L. PLoS One 10:e0129696

Romanowska E, Wróblewska B, Drożak A, Zienkiewicz M, Siedlecka M (2008) Effect of Pb ions on superoxide dismutase and catalase activities in leaves of pea plants grown in high and low irradiance. Biol Planta 52:80

Romanowska E, Wasilewska W, Fristedt R, Vener AV, Zienkiewicz M (2012) Phosphorylation of PSII proteins in maize thylakoids in the presence of Pb ions. J Plant Physiol 169:345–352

Rouhier N, Cerveau D, Couturier J, Reichheld JP, Rey P (2015) Involvement of thiol-based mechanisms in plant development. BBA-Gen Subj 1850:1479–1496

Ruban AV, Johnson MP (2009) Dynamics of higher plant photosystem cross-section associated with state transitions. Photosynth Res 99:173–183

Schultz C, Soll J, Fiedler E, Schultze-Siebert D (1985) Synthesis of prenylquinones in chloroplasts. Physiol Planta 64:123–129

Sersen F, Kralova K, Pesko M, Cigan M (2014) Effect of Pb^{2+} ions on photosynthetic apparatus. Gen Physiol Biophys 33:131–136

Shahid M, Pourrut B, Dumat C, Nadeem M, Aslam M, Pinelli E (2014) Heavy metal induced reactive oxygen species: phytotoxicity and physicochemical changes in plants. In: Whitacre DM (ed) Reviews of environmental contamination and toxicology, vol 232. Springer, Cham, pp 114–131

Shahid M, Dumat C, Pourrut B, Abbas G, Shahid N, Pinelli E (2015) Role of metal speciation in lead-induced oxidative stress to *Vicia faba* roots. Russ J Plant Physiol 62:448–454

Shahid M, Dumat C, Khalid S, Schreck E, Xiong T, Niazi NK (2017) Foliar heavy metal uptake, toxicity and detoxification in plants: a comparison of foliar and root metal uptake. J Hazard Mater 325:36–58

Sharma P, Dubey RS (2005) Lead toxicity in plants. Braz J Plant Physiol 17:35–52

Shen J, Song L, Müller K, Hu Y, Song Y, Yu W, Wang H, Wu J (2016) Magnesium alleviates adverse effects of lead on growth, photosynthesis, and ultrastructural alterations of *Torreya grandis* seedlings. Front Plant Sci 7:1819

Shi X, Wang S, Wang D, Sun H, Chen Y, Liu J, Jiang Z (2019) Woody species *Rhus chinensis* Mill. seedlings tolerance to Pb: physiological and biochemical response. J Environ Sci 78:63–73

Singh R, Tripathi RD, Dwivedi S, Kumar A, Trivedi PK, Chakrabarty D (2010) Lead bioaccumulation potential of an aquatic macrophyte *Najas indica* are related to antioxidant system. Bioresour Technol 101:3025–3032

Slootweg J, Hettelingh J, Posch M (2010) Critical loads of heavy metals and their exceedances. In: Slootweg J, Posch M, Hettelingh JP (eds) Progress in the modelling of critical thresholds and dynamic modelling, including impacts on vegetation in Europe, CCE Status Report No 680359001. Coordination Centre for Effects, Bilthoven, Netherlands, pp 91–100

Sobotik M, Ivanov VB, Obroucheva NV, Seregin IV, Martin ML, Antipova OV, Bergmann H (1998) Barrier role of root system in lead-exposed plants. Angew Bot 72:144–147

Stefanov D, Ananieva K, Gesheva E, Dinev N, Nikova I, Vasilev A, Doncheva S (2014) PTOX-dependent electron flow is involved in photosynthetic electron transport protection against heavy metal toxicity in sunflower leaves. Comp Rend Acad Sci 67:931–936

Sun T, Yuan H, Cao H, Yazdani M, Tadmor Y, Li L (2018) Carotenoid metabolism in plants: the role of plastids. Mol Plant 11:58–74

Tyystjärvi E (2008) Photoinhibition of photosystem II and photodamage of the oxygen evolving manganese cluster. Coord Chem Rev 252:361–376

Wasilewska W, Drożak A, Bacławska I, Kąkol K, Romanowska E (2015) Lead induced changes in phosphorylation of PSII proteins in low light grown pea plants. Biometals 28:151–162

Wood WH, MacGregor-Chatwin C, Barnett SF, Mayneord GE, Huang X, Hobbs JK, Hunter CN, Johnson MP (2018) Dynamic thylakoid stacking regulates the balance between linear and cyclic photosynthetic electron transfer. Nat Plant 4:116–127

Xiong ZT, Zhao F, Li MJ (2006) Lead toxicity in *Brassica pekinensis* Rupr.: effect on nitrate assimilation and growth. Environ Toxicol 21:147–153

Zhong B, Chen J, Shafi M, Guo J, Wang Y, Wu J, Ye Z, He L, Liu D (2017) Effect of lead (Pb) on antioxidation system and accumulation ability of Moso bamboo (*Phyllostachys pubescens*). Ecotox Environ Safe 138:71–77

Zhou C, Huang M, Li Y, Luo J, Cai L (2016) Changes in subcellular distribution and antioxidant compounds involved in Pb accumulation and detoxification in *Neyraudia reynaudiana*. Environ Sci Pollut Res 23:21794–21804

Zhou J, Jiang Z, Ma J, Yang L, Wei Y (2017) The effects of lead stress on photosynthetic function and chloroplast ultrastructure of *Robinia pseudoacacia* seedlings. Environ Sci Pollut Res 24:10718–10726

Physiological and Biochemical Changes in Plant Growth and Different Plant Enzymes in Response to Lead Stress

Eda Dalyan, Elif Yüzbaşıoğlu, and Ilgın Akpınar

Abstract Lead (Pb) is one of the most widespread, persistent and toxic heavy metal contaminants in agricultural soil. Though Pb is not an essential metal for plant metabolism, it is taken up primarily by the root system and accumulated in the different plant parts. Because Pb ions accumulate predominantly in roots, root growth is more sensitive to this metal than shoot growth. Growth inhibition due to Pb stress depends on various mechanisms affecting directly (such as reduction of cell division and elongation) or indirectly (such as disorders nutrient uptake, photosynthesis and water uptake) plant growth. After entering the cell, Pb ions can also influence the activity of the key enzymes of different metabolic processes such as antioxidative and photosynthesis. Pb stress might inhibit or induce the activity of these enzymes depending on the plant species, metal type and concentration, and duration of the exposure. The inhibition of enzyme activity by Pb mostly arises from the interaction between the Pb and enzyme sulfhydryl groups. Also, inhibition of metalloenzymes under Pb stress may occur due to the displacement of an essential metal by Pb ion. Furthermore, activities of certain enzymes induced by Pb stress might result from the changes in enzyme synthesis, immobilization of their inhibitors. This chapter reviews from the point of view of physiological and biochemical mechanisms the alterations occurring in growth and the activations of different enzymes in plants due to Pb stress.

Keywords Lead · Growth · Cell division · Cell elongation · Enzyme activity

E. Dalyan (✉) · E. Yüzbaşıoğlu
Faculty of Science, Department of Botany, Istanbul University, Istanbul, Turkey
e-mail: ekaplan@istanbul.edu.tr

I. Akpınar
Institute of Sciences, Istanbul University, Istanbul, Turkey

© Springer Nature Switzerland AG 2020
D. K. Gupta et al. (eds.), *Lead in Plants and the Environment*, Radionuclides
and Heavy Metals in the Environment,
https://doi.org/10.1007/978-3-030-21638-2_8

1   Introduction

Lead (Pb) is one of the most dangerous heavy metal contaminants of the environment. It is ranked second among all the hazardous substances due to its toxicity, occurrence, and distribution over the globe (ATSDR 2017). Apart from the occurrence of natural, the concentration of Pb is increasing rapidly in the environment with the growth of industrialization and human activities. Mining and smelting activities, automobile exhausts, coal burning, effluents from storage battery industries, lead-containing paints, paper and pulp, pesticides, as well as the disposal of municipal sewage sludge are the main sources of Pb contamination (Khan et al. 2018a). Pb is readily absorbed by plants from the contaminated soil, water and atmosphere cause hazardous health effects on humans and animals via the food chain (Rizwan et al. 2018).

Lead is a non-essential element for plants and has no biological function but several plant species grow on Pb contaminated soils. Pb is mostly absorbed by the plant root system and its small quantity is translocated to the above-ground parts of the plant. Since the majority of Pb absorbed by the plant is accumulated in their roots (approximately 95% or more), Pb mainly affects plants via their root system (Dalyan et al. 2018). Pb stress either directly or indirectly causes severe alteration into various physiological and biochemical processes such as growth, photosynthesis, respiration, mineral nutrition, water uptake, and enzyme activities in plants. Responses of plants to Pb stress include disturbance in mitosis, damage of DNA synthesis (Kumar et al. 2017), inhibition of root and shoot growth (Chen et al. 2017), disruption of mineral nutrition, inhibition of photosynthetic pigments, reduction in photosynthesis, transpiration, and water uptake (Jayasri and Suthindhiran 2016), inhibition or activation of enzymatic activities (Sidhu et al. 2016). However, the effects of Pb stress on the plant can alter depending on the metal type and concentration, exposure time, plant species, the stages of plant development, and different plant organs (Pourrut et al. 2011).

2   Effects of Lead Stress on Growth

Pb stress negatively affects growth and biomass in plants. Pb in growth medium generally causes a decrease in growth parameters such as elongation, fresh and dry biomass in plant root and shoot. For instance, *Medicago sativa* seedlings exposed to Pb(NO$_3$)$_2$ (0, 10 and 100 µM) for 7 days were gradually decreased in the lengths and fresh weights of root and shoot (Hattab et al. 2016). Also, the different concentrations (0, 100, 200, 300, 400 and 500 mg L^{-1}) of Pb for 12 days in the *Acalypha indica* caused an adverse effect on growth index by reducing the length and fresh and dry biomass of the root and shoot (Venkatachalam et al. 2017). Some other studies reported similar results: root, shoot and leaf growth, fresh and dry biomass were severely decreased by Pb stress in *Triticum aestivum* (Kaur et al. 2012a), *Zea mays* L. (Hussain et al. 2013), cotton (Bharwana et al. 2013), *Sesbania grandiflora*

(Malar et al. 2014a), *Brassica napus* L. (Shakoor et al. 2014), *Brassica juncea* L. (Kohli et al. 2018), and *Ricinus communis* L. and *Brachiaria mutica* (Khan et al. 2018b). Furthermore, the effects of Pb stress on plant growth might show alteration depending on exposure time, metal concentration and plant growth stages. A study with *Zea mays* seedlings exposure to Pb (0, 25, 50, 100, 200 µM) for 1–4–7 days, both root and shoot dry weights did not change at all Pb treatments, with exception after 7 days where shoot dry weight decreased at 200 µM Pb. In addition, shoot fresh weight and root length were not changed in any of Pb treatments (Gupta et al. 2009). Similarly, the different Pb concentrations did not exhibit remarkable effects on the growth parameters in lettuce (Silva et al. 2017).

The growth inhibition due to Pb stress is stronger in the plant roots. Since the most of the Pb taken up by plants is accumulated in the roots (95%) and only a small amount (5%) is transported to the above-ground parts, root growth is more influenced by Pb toxicity (Zhou et al. 2016). In *Triticum aestivum* exposed to different Pb concentrations (500, 1000, 2500 µM) for 7 days, root length reduced in the range of ~23–51% over 500–2500 µM Pb whereas shoot length was decreased by ~17%, 31% and 44% (Kaur et al. 2012a). Similarly, Pb treatment leads to ~67% reduction in root growth of two maize varieties though shoot length remained less affected (Ghani et al. 2010). Furthermore, it was reported that the length and biomass of root decreased due to Pb stress in several plants including *Lathyrus sativus* (Brunet et al. 2008), radish (Gopal and Rizvi 2008), *Zea mays* L. (Kozhevnikova et al. 2009), *Sedum alfredii* (Gupta et al. 2010). This inhibition of root growth might be originated from disturbances in either cell division and/or cell elongation (Rucińska et al. 1999; Kozhevnikova et al. 2009). The decrease of the cell division and elongation rates in the plant roots is associated with various mechanisms including change of cell wall plasticity, inhibition of microtubule development and DNA synthesis, metal-induced chromosomal aberrations, expansion of the mitotic cycle, and reduction of glutathione pool (Seregin and Ivanov 2001).

The antimitotic effect is one of the best-known toxic effects of Pb stress on plants (Shahid et al. 2011). In the first mechanism, Pb ions taken up by roots might bind to cell wall and the cell membranes and thus lead to cell wall mineralization known as calcification and silicification (Wierzbicka 1998). Since cell wall mineralization produces rigidity in the components of the cell wall, the physical and chemical characterization of cell wall change. These alterations in cell wall plasticity cause a reduction in cell division or elongation. In the second mechanism, microtubules that are the main element for mitosis are disrupted. Pb ions negatively affect the G2 and M phases of cell division and lead to the formation of abnormal cells at the colchicine-mitosis stage. This event might be related to direct or indirect interactions between Pb ions and proteins such as cyclins, whose activity is indirectly dependent on glutathione concentration, involved in the cell cycle. The spindle activity disorders due to Pb stress can be temporarily seen and return to the mitotic index initial levels (Shahid et al. 2011).

Pb stress causes a reduction in the number of dividing cells by affecting the normal cell cycle. The decrease of mitotic activity possibly due to blocking of G2 phase prevents the cell from entering the mitosis (Sudhakar and Venu 2001). This effect might be due to the inhibition of microtubule formation and DNA synthesis, degradation of nucleoprotein synthesis and decreased the ATP level to supply energy to microtubule dynamics, spindle elongation, and chromosomal movement (Türkoğlu 2012). Pb stress induced the DNA damage leads to the single and double strand breaks of DNA and irreversible in DNA replication, transcription, and repair. Pb stress was reported to increase the DNA damage in root cells of *Lupinus luteus* (Rucińska et al. 2004), *Nicotiana tabacum* (Gichner et al. 2008), and *Solanum lycopersicum* (Kumar et al. 2017) depending on concentration and exposure time. The disorders of cell division and elongation might occur as a result of direct binding of metal to DNA, disruption of microtubule organization and suppression of cytokines (Seregin and Kozhevnikova 2006).

Although the low concentration of Pb does not have an important effect on mitosis, it induces chromosomal aberrations such as of chromosome bridge formations during the anaphase, chromosome fragmentation, eccentric fragment loss during meiosis, and micronuclei formation (Shahid et al. 2011; Rodriguez et al. 2013). The chromosomal aberrations induced by Pb can be explicated by the negative effects of Pb on the mitotic spindle, microtubule formation and DNA synthesis. Colchicine mitosis and chromosome stickiness are the major chromosomal aberrations. The increased cells with colchicine mitosis are toxic properties of low Pb concentration. Pb behaves in a similar manner to colchicine, a cell division inhibitor. Colchicine inhibits microtubule polymerization, and causes the formation of characteristic colchicine-mitosis by blocking the cells in prometaphase (Checchi et al. 2003). Also, the increment of chromosome stickiness which is another chromosomal aberration is associated with DNA damage depending on Pb stress (Jiang et al. 2014).

In the roots of *Allium cepa* exposed to Pb stress, Pb ions in root cells disorganized the network of microtubules and caused in an important reduction in the chromosomal movement and segregation (Jiang et al. 2014). The possible interplay between Pb and tubulin inhibits the polymerization, damages to the mitotic spindle fibers, and thus leads to increase in colchicine-mitosis stage and chromosomal stickiness, and decrease in mitotic index (Jiang et al. 2014). Similarly, in another study, mitotic index together with the mitotic abnormalities such as metaphase and telophase, vagrant, laggards, and sticky chromosomes were inhibited in onion root tip exposed to Pb (16.6–331 mg L^{-1}) depending on metal concentration (Kaur et al. 2014).

Plant growth inhibition due to Pb stress might be associated with several mechanisms including disrupting nutrient uptake, disturbing photosynthesis, lowering of water potential and increasing oxidative stress (Riffat et al. 2009).

2.1 Mineral Nutrition

Balanced mineral nutrient supply is very important for plant growth. The macro- and micronutrients necessary for the normal growth of plants can alleviate the negative effects of different environmental stresses by developing the physiological and molecular mechanisms of plants (Arshad et al. 2016). These essential nutrients are the main components of many metabolic active compounds that regulate different physiological functions. For instance; nitrogen, phosphorus, potassium, calcium and magnesium are directly or indirectly very important for cell division, cell expansion and differentiation (Alamri et al. 2018).

Pb stress disrupts the plant and nutrient relationship and alters the ratios of internal nutrient between plant tissues (Gopal and Rizvi 2008). The reduction in the growth and biomass of plant due to metal stress originates from alterations in diverse biochemical processes at the cellular level affecting the nutrient uptake and metabolism (Ali et al. 2015). Generally, there is a negative correlation between the contents of mineral nutrient and Pb stress in plants. It was reported that the concentration of some macro- and micronutrients decreased in wheat plants treated with the different concentrations (0, 1.5, 3 and 15 mM) of Pb (Lamhamdi et al. 2013). Similar results were exhibited with the reduction of calcium, magnesium, sodium and potassium concentrations in the shoot and roots of maize plant exposed to Pb stress (Singh et al. 2015). It was also found that Pb treatment led to decrease the concentration of zinc, iron, manganese, copper, calcium, phosphorus and magnesium in *Oryza sativa* (Chatterjee et al. 2004), *Medicago sativa* (Lopez et al. 2007), and *Raphanus sativus* (Gopal and Rizvi 2008).

The reduced uptake of mineral nutrients might occur with two different mechanisms. First mechanism is related to the metal ions size. The competition between the metal ions which has similar size such as potassium ions with Pb might lead to the reduced mineral uptake. Since the interaction between these two metals which are similar radii (Pb^{2+}: 1.29 Å and K^+: 1.33 Å) is strong, these ions might compete for entry into the plant through the same potassium channels (Sharma and Dubey 2005). Furthermore, phosphorus shows a negative correlation with Pb amounts in soil (Päivöke 2002). Second metabolism might originate from alterations occurring in cell metabolism in response to metal stress by disturbing cell membrane and inhibiting enzymatic activities. Although Pb stress decreases nitrate uptake from the soil, it does not affect the nitrogen flow of the plant cell. This reduction in nitrogen amounts can be induced by decrease in the activity of nitrate reductase, which acts as rate-limiting in nitrate assimilation (Xiong et al. 2006). Tariq and Rashid (2013) showed a significant reduction in nitrate amount and the activity of nitrate reductase in rice seedlings growing in Pb-contaminated soils. Also, Burzyński and Grabowski (1984) exhibited that decreased nitrate uptake might originate from moisture stress depending on Pb stress.

2.2 Photosynthesis

Pb stress causes membrane content and permeability change, deterioration of the chloroplast ultrastructure organization, disruption of organs including chlorophyll, plastoquinone and carotenoid, and inhibition of enzymes in the Calvin cycle. It also leads to the destruction of photosynthetic pigments such as chlorophyll or the inhibition of the synthesis of these pigments. As a result of these cellular processes, CO_2 deficiency and resulting stomata closure further affect photosynthesis (Khan et al. 2018b).

Pb-induced the reduction of total chlorophyll content might be due to increased chlorophyll degradation, deterioration of chloroplast stromal volume (Stefanov et al. 1995; Hadi and Aziz 2015), or impairment the uptake of main photosynthetic pigment elements, such as magnesium, iron, calcium and potassium (Piotrowska et al. 2009). A reduction in total chlorophyll amount was determined in water hyacinth applied different Pb concentrations (0, 100, 200, 400, 600, 800 and 1000 mg L^{-1}) (Malar et al. 2014b). Similarly, decrease in the level of photosynthetic pigments including chlorophyll a, b and carotenoids was observed in many plant species such as ryegrass, cotton and rice (Bai et al. 2015; Khan et al. 2016; Chen et al. 2017). However, there are also few studies showing that Pb has a positive effect on chlorophyll pigments (Mroczek-Zdyrska et al. 2017). Pb stress generally affects chlorophyll a more than chlorophyll b (Hou et al. 2018). Whereas, Malar et al. (2014b) observed that the most chlorophyll b among photosynthetic pigments was reduced in water hyacinth exposed to Pb. This might be attributed to the change in the photosynthetic pigment composition comprising lower level of light harvesting chlorophyll proteins (LHCPS) (Malar et al. 2014b).

Chloroplast pigments are the main components of photosynthesis, and are responsible for plant biomass production. Pb stress significantly reduces plant growth by decreasing total chlorophyll levels (Aliu et al. 2013). For instance, dry biomass and total chlorophyll amounts showed a reduction in a species of algae under Pb stress. The decrease in dry biomass of the plant might be due to inhibition of cell division which causes growth reduction (Choudhury and Panda 2005). Similarly, it was reported that Pb stress led to a decrease in dry weight and chlorophyll content in lettuce (Đurđević et al. 2008) and wheat (Kaur et al. 2012b). In addition, plant growth and overall biomass production were determined in *Lathyrus sativus* plants exposed to Pb. It has been revealed that Pb can directly affect elongation by inhibiting cell wall enzymes and the plasmalemma ATPase by damaging electron transport in the process of photosynthesis, and thus might cause growth inhibition (Abdelkrim et al. 2018).

2.3   Water Uptake

The growth and physiological mechanisms in plants are directly or indirectly regulated by the water supply. Pb stress leads to decrease in transpiration rate and water content in plants. These negative effects depend on various mechanisms. First, growth retardation occurs with the decreased of leaf area, the most important transpiration organ (Brunet et al. 2009). Second, guard cells are smaller in plants exposed to heavy metals because heavy metals affect leaf growth more than stomata differentiation (Weryszko-Chmielewska and Chwil 2005). Third, Pb ions decrease the water potential by reducing the content of compounds that protect cell turgor and cell wall plasticity such as sugars and amino acids, and this effect is the most important factor of growth inhibition (Barceló and Poschenrieder 1990). For instance, 2 mM Pb treatment was reduced relative water content in wheat plants, and this effect might be due to the growth inhibition or the alterations of cell wall extensibility and cell wall elasticity (Alamri et al. 2018). Fourth, Pb causes the stomata to close by increasing the content of ABA (Atici et al. 2005; Weryszko-Chmielewska and Chwil 2005) and significantly restricts the gas flow between the leaves and the atmosphere. This reduction in incoming CO_2 flow is thought to be the major cause of a significant decrease in CO_2 fixation. With the same interaction mechanisms, Pb might also inhibit some enzymes from Calvin cycle (Romanowska et al. 2002).

3   Effects of Lead Stress on Different Enzyme Activities

Pb stress causes the alteration in many physiological processes by leading to inhibition or induction of some enzymatic activities. Pb has an important effect on enzyme activity, but the mechanism of this interaction has still ambiguous. Generally, Pb ions directly inhibit the enzyme activity due to change the inactivation constant (Ki) of enzymes, and high affinity of –SH and –COOH groups on the enzymes (Seregin and Ivanov 2001; Gupta et al. 2009, 2010). Also, Pb ions can replace the other divalent cations which are necessary for enzyme activation, such as zinc, manganese, and iron or inhibit the absorption of these minerals (Seregin and Ivanov 2001). Furthermore, the indirect effect of Pb on enzymes results from an increase of reactive oxygen species which leads to the oxidative damage on proteins in plants (Kumar and Prasad 2018).

3.1   Reactive Oxygen Species Production and Lipid Peroxidation

Pb stress primarily causes oxidative stress by increasing reactive oxygen species (ROS) production in plants cell (Gill and Tuteja 2010). Many studies reported that Pb caused to the overproduction of ROS such as singlet oxygen ($^1O_2$), superoxide (O^{2-}), hydroxyl radical ($HO^{\bullet}$), and hydrogen peroxide (H_2O_2) in chloroplast, mitochondria and peroxisomes in various plant species (Liu et al. 2008; Hattab et al. 2016). The excessive accumulation of ROS inhibits normal cell functions such as photosynthetic activity, ATP production, and protein synthesis. In addition, overproduction of ROS damages cell and organelle membrane, photosynthetic pigments and nucleic acid (Gill and Tuteja 2010). Pb also causes the leakage of potassium with the alteration in the composition of the lipid bilayer, and leads to the lipid peroxidation in the cell membrane structure. As a result of lipid peroxidation, the contents of saturated fatty acid reduce, while the contents of unsaturated fatty acid enhance in the cell membrane of many plant species under Pb toxicity (Kumar and Prasad 2018). These change in lipid composition of the membrane cause abnormalities of cellular structures such as cytoskeleton structure and ultrastructure of organelles (Verma and Dubey 2003).

There is been a correlation between lipid peroxidation and ROS production. Lipid peroxidation is commonly known as a biochemical indicator for ROS injury in plants (Gill and Tuteja 2010). However, ROS production does not responsible for the inducement of lipid peroxidation under Pb stress. The metal ions might trigger the free radical production which is originated by apoplastic and enzymatic (Mika et al. 2004). The role of membrane NADPH-oxidases (NOX) in the response to heavy metals such as copper, nickel and cadmium in plants was revealed in many studies (Quartacci et al. 2001; Olmos et al. 2003; Hao et al. 2006). Pourrut (2008) firstly proved that NOX activity was the primary source of oxidative burst under Pb stress. In addition, lipoxygenases (LOX) and phospholipases can enzymatically catalyze the lipid peroxidation (Huang et al. 2012). Pb ions upregulate the gene expression of lipoxygenases and stimulate LOX enzyme activity (Huang et al. 2012). Kaur et al. (2012b) showed that malondialdehyde (MDA) and H_2O_2 content increased in wheat root exposed to 500, 1000, and 2500 µM Pb. Similarly, it was shown that MDA content increased in *Zea mays* L. (Gupta et al. 2009), *Triticum aestivum* (Yang et al. 2011) and *Oryza sativa* (Thakur et al. 2017) under Pb stress. Mroczek-Zdyrska et al. (2017) reported that H_2O_2 and $O_2^{\cdot-}$ accumulation significantly enhanced in 50 mM Pb-treated *Vicia* plants by performing the histochemical analysis of oxidative stress. Another study revealed that the Pb stress increased the free radical (H_2O_2 and $O_2^{\cdot-}$) generation and MDA accumulation in roots and leaves of both *Arachis hypogaea* L. cultivars (Nareshkumar et al. 2015).

3.2 Antioxidative Mechanism

Plants can develop antioxidant enzyme systems for scavenging excessive accumulation of ROS under metal stress. The enzymatic antioxidants includes the key enzymes such as superoxide dismutase (SOD), catalase (CAT), peroxidases (POX), guaiacol peroxidase (GPX), ascorbate peroxidase (APX), glutathione reductase (GR), glutathione S-transferases (GST) (Hattab et al. 2016; Alamri et al. 2018; Khan et al. 2018b). When the plant cell produces the excessive ROS as a consequence of Pb toxicity, SOD, a metalloenzyme, is the first defense enzyme that converts $O_2^{\cdot-}$ radicals to form H_2O_2 and O_2 (Hasanuzzaman et al. 2012). H_2O_2 is a very reactive strong oxidant because of being unpaired electrons and is removed by activities of CAT that decomposes H_2O_2 to water and molecular oxygen (Gill and Tuteja 2010). Also, peroxidases are a potential antioxidant for removal of H_2O_2 molecules by turns it into H_2O and O_2 (Singh et al. 2010). Furthermore, the other H_2O_2-scavenging enzyme system occur the Halliwell–Asada pathway including ascorbate–glutathione cycle enzymes such as ascorbate peroxidase, dehydroascorbate reductase (DHAR), monodehydroascorbate reductase (MDHAR) and glutathione reductase (GR) (Potters et al. 2010). Firstly, APX catalyzes the reduction of H_2O_2 into monodehydroascorbate radical (MDHA) in the presence of ascorbate. MDHA enzymatically is converted to AsA and DHA by the activity of MDHAR (Małecka et al. 2009; Gill and Tuteja 2010). DHAR converts DHA to ascorbate using glutathione as an electron donor, for the glutathione disulfide (GSSG) production. Finally, GR enzyme catalyzes GSSG to glutathione by using NADPH-dependent reduction (Potters et al. 2010). In addition, GST enzymes are also important in ROS and metal detoxification and abiotic stress tolerance (Gill and Tuteja 2010; Hasanuzzaman et al. 2012).

Several studies have determined that Pb stress can inhibit or induce the activity of antioxidant enzymes depending on the duration or concentration of the treatment and plant species (Table 1) (Islam et al. 2008; Singh et al. 2010). It was studied the changes of antioxidant enzyme activities in *Acalypha indica* exposed to lead (100–500 mg L^{-1}) for 1–12 days under hydroponic culture (Venkatachalam et al. 2017). This study revealed that Pb-treated plants significantly enhanced the SOD, CAT, APX and POX activities which diminish the metal-induced phytotoxicity (removal of excess ROS) (Venkatachalam et al. 2017). Gupta et al. (2009) were investigated how maize adapted to the different Pb concentrations (0–200 μM) for 1–7 days. Their results showed that antioxidant enzymes such as SOD and CAT, as well as ascorbic acid level, alleviated Pb stress, which enhanced linearly with increasing Pb concentrations and exposure time (Gupta et al. 2009). In *Arachis hypogaea* L. cultivars, antioxidant enzymes such as SOD, APX, GPX, GR and GST exhibited increase under Pb stress, and also the isozyme band intensities of SOD, APX and GPX were consistent with alterations in the activities of antioxidative enzyme (Nareshkumar et al. 2015). Pb stress led to decrease in ascorbate content by reducing MDHAR and DHAR activities in wheat seedling (Hasanuzzaman et al. 2018). Dalyan et al. (2018) showed that Pb toxicity (2 mM) stimulated the increase of SOD, CAT, POX, APOX, GR and GST enzyme activities in the root of *Brassica juncea* seedlings. In another study, it was revealed that different Pb levels caused more increase in ascorbic acid and H$_2$O$_2$ contents, as well as DHA, CAT and APX activities in the roots of metal accumulator *S. alfredii* than non-accumulator (Huang et al. 2012). Although most studies have emphasized important of antioxidant enzymes in response to Pb stress, also it has been reported studies showing changes in the activation of other oxidant enzymes. For instance, the activity of aldo–keto reductase which detoxifies the aldehydes and associated with detoxification of Pb-induced toxicity increased in *T. triangulare* roots under Pb stress (Kumar and Majeti 2014; Ashraf et al. 2015). Additionally, Bali et al. (2019) determined that polyphenol oxidase (PPO) activity increased at 0.25–0.75 mM Pb concentration in tomato seedlings.

3.3   Photosynthesis

Photosynthesis is one of the most important processes that are negatively affected by Pb stress in plants. The Pb toxicity causes to damage in chloroplast membrane and ultrastructure, inhibition of chlorophyll, plastoquinone and carotenoid synthesis and the reduction of enzymatic activities in the Calvin cycle (Stefanov et al. 1995; Liu et al. 2008). It also leads to destruction in the lipid composition of thylakoid membranes in the chloroplasts (Stefanov et al. 1995). Initially, Pb-exposed plants show a

Table 1 List of antioxidant enzymes in various plant species exposed to different concentrations of Pb

Plant species	Pb concentration	Antioxidant enzymes	References
Gossypium spp.	50–100 µM	SOD, GPX, APX, CAT	Bharwana et al. (2013)
Sesbania drummondii	500 mg L⁻¹	SOD, GPX, APX, CAT	Ruley et al. (2004)
Pisum sativum	1 mM	Cu, Zn-SOD, Mn-SOD, CAT	Małecka et al. (2009)
Triticum aestivum	0.5, 1, 2 mM	APX, MDHAR, DHAR, GR, SOD, CAT, GR, GPX, GST	Hasanuzzaman et al. (2018); Alamri et al. (2018)
Brassica juncea	50–500 mg kg⁻¹	SOD, CAT, APX	John et al. (2009)
Raphanus sativus	25–500 ppm, 0.1, 0.5 mM	CAT and POD isoenzymes, APX, GPX, POX, acid phosphatase, ribonuclease	Gopal and Rizvi (2008); El-Beltagi and Mohamed (2010)
Oryza sativa	1 mM	SOD, CAT, POX	Khan et al. (2018a)
Brassica oleracea L. convar. botrytis	0.25, 0.5 mM	SOD, CAT, POD, APX, GR	Chen et al. (2018)
Zea mays	25–200 µM, 1–20 mM	SOD, CAT	Gupta et al. (2009); Chen et al. (2018)
Brassica juncea	0.25, 0.50, 0.75 mM	POD, APOX, GR, DHAR, MDHAR, GST, GPOX, SOD, CAT, PPO	Kohli et al. (2018)
Pogonatherum crinitum	500–2500 mg kg⁻¹	POD, CAT	Hou et al. (2018)
Brachiaria mutica	100–500 µM	SOD, GPOD, CAT	Khan et al. (2018b)
Lathyrus sativus	0.5 mM	GPOX, SOD, APX, CAT	Abdelkrim et al. (2018)
Cassia angustifolia	100–500 µM	SOD, APX, GR, CAT	Qureshi et al. (2007)
Macrotyloma uniflorum *Cicer arietinum*	200–800 ppm	SOD, CAT, POD, GR, GST	Reddy et al. (2005)
Eichhornia crassipes	1000 mg L⁻¹	APX, POX, CAT, SOD	Malar et al. (2014b)
Ceratophyllum demersum	1–100 µM	SOD, GPX, APX, CAT, GR	Mishra et al. (2006)
Lycopersicon lycopersicum	0.25–0.75 mM	SOD, POD, CAT, GPOX, GST, GR, PPO	Bali et al. (2019)
Ricinus communis	100–500 µM	SOD, GPOD, CAT	Khan et al. (2018b)

dramatic decrease in pigment content due to increased chlorophyll degradation. The degradation process of chlorophyll pigments can be explained by increasing chlorophyllase, protochlorophyllide reductase, pheophorbide oxygenase, Mg-dechelatase, and red chlorophyll catabolite reductase activities (Drazkiewicz 1994; Harpaz-Saad et al. 2007; Liu et al. 2008). Moreover, Pb ions accelerate chlorophyll degradation by reduced uptake of the essential components of chlorophyll structure such as Mg and Fe (Burzynski 1987). In addition, Pb disturbs the structure of metalloenzymes such as δ-aminolevulinic acid dehydratase (ALAD) which is an essential metalloenzyme in chlorophyll biosynthesis by replacing with the divalent ions (Cenkci et al. 2010). Under Pb toxicity, Zn^{2+} ion of ALAD enzyme is replaced by Pb^{2+} (Cenkci et al. 2010). Cenkci et al. (2010) determined that 0.5–5 mM concentrations of lead nitrate decreased photosynthetic pigment content due to inhibit ALAD activity in 20-day-old seedlings of *Brassica rapa*. On the other hand, Pb ions led to reduction in the activities of ferredoxin NADP+ reductase (Gupta et al. 2009); inhibition activities of Calvin cycle enzymes (Xiong et al. 2006; Singh et al. 2010); decrease the activity of phosphoenolpyruvate carboxylase (PEPC) (Sharma and Dubey 2005); decline of the rubisco activity (Alamri et al. 2018) in plant photosynthetic process. Alamri et al. (2018) revealed that 2 mM Pb treatment significantly reduced the Rubisco activity in wheat leaves. Many researchers reported that Pb stress led to loss of the photosynthetic pigments including chlorophyll a, chlorophyll b and total carotenoids in radish (El-Beltagi and Mohamed 2010), *Brassica rapa* (Cenkci et al. 2010), wheat and spinach (Lamhamdi et al. 2013), *Jatropha curcas* (Shu et al. 2012), maize (Zhang et al. 2018), *Pogonatherum crinitum* (Hou et al. 2018).

3.4   Respiration

The excessive accumulation of Pb has a detrimental effect on plant physiology and biochemistry. Especially, lead toxicity has harmful effect on mitochondrial membranes and respiratory enzymes in plants. Lead stress causes a decrease in respiration rate via inhibiting flow the electrons in the electron transport system (Bazzaz et al. 1975; Ashraf et al. 2015). Mostly, Pb ions damage the electron transport chain reaction in result binding to mitochondrial membranes. It was exhibited that Pb stress inhibited ATP synthetase/ATPase activity in maize (Tu and Brouillette 1987). However, Romanowska et al. (2002) determined that 5 mM lead nitrate increased ATP content and the ATP/ADP ratio in detached leaves of pea and maize in respiration for 24 h.

3.5   Nitrate Assimilation

Pb toxicity disturbs the balance of inorganic compounds by blocking of nitrate uptake or inhibiting activity of nitrate reductase (NR) in the plants (Sengar et al. 2009). NR, a metalloprotein, is the first enzyme in the nitrogen assimilation pathway that converts nitrate to nitrite in the cytosol (Ashraf et al. 2015). Also, NR catalyzes nitric oxide synthesis which is a signal molecule in plants (Khan et al. 2017; Alamri et al. 2018). Many studies reported that lead toxicity caused a reduction in the activity of nitrate reductase and limit the nitrate assimilation in plants (Xiong et al. 2006; Sengar et al. 2009). Pb stress led to the significant decrement of nitrate reductase activity in rice seedlings (Tariq and Rashid 2013; Sharma and Dubey 2005). Furthermore, Xiong et al. (2006)

showed that NR activity significantly decreased in *B. pekinensis* treated with lead (4 and 8 mmol kg⁻¹). In addition, NR activity negatively affected under Pb treatment in *Pisum sativum*, *Medicago sativa*, *Picea rubens* and *Triticum aestivum* (Porter and Sheridan 1981; Päıvöke 1983; Yandow and Klein 1986; Alamri et al. 2018).

3.6 Soluble Protein

High concentration of Pb not only changes the activity of enzyme but also it alters the soluble protein content in plants (Mishra et al. 2006; Singh et al. 2010). Shu et al. (2012) determined that high concentration of Pb ions reduced the protein content in *Jatropha curcas* L. Although the low concentrations of Pb increase protein content by stimulating stress proteins, the high concentrations of Pb trigger to catabolism of protein synthesis. This reduction can be explained by proteins oxidation (Wang et al. 2008), the increase of ribonuclease activity, and the decrease of protease activity (Gopal and Rizvi 2008). Excessive Pb treatment increased the ribonuclease activity in radish (Gopal and Rizvi 2008), rice (Mukherji and Maitra 1976) and maize (Maier 1978).

3.7 Proline Synthesis

One of the important defense strategies against Pb is biosynthesis of low molecular weight proteins such as proline (Rucińska-Sobkowiak et al. 2013). Proline, is known as an organic osmolyte, plays an important role for scavenging ROS and protecting the protein structure in plants under stress conditions (Parys et al. 2014). Pb stress controls proline accumulation by activating ornithine-δ-aminotransferase and r-glutamyl kinase enzymes in ornithine and glutamate biosynthesis pathways (Yang et al. 2011; Kumar and Prasad 2018). A study with wheat cultivar showed that short term Pb stress (1–4 mM) led to increased accumulation of proline by induction of ornithine-δ-aminotransferase and r-glutamyl kinase enzymes (Yang et al. 2011).

3.8 Seed Germination

Lead has an inhibitory effect in the seed germination process of many plant species (Pandey et al. 2007). The major effect of Pb is seen on the inhibition of germination enzymes such as protease and amylase which is synthesized during the germination. Pandey et al. (2007) reported that amylase and protease activities decreased seed germination of *Catharanthus roseus* under high concentration of Pb. Lamhamdi et al. (2011) observed that amylase activity in wheat seeds was reduced with increasing level of Pb. This inhibition might show that lead ions replace the calcium ions which are essential for the activities of these enzymes (Lamhamdi et al. 2011).

3.9 Phytochelatins

Pb ions induce the synthesis of phytochelatins (PCs) that has an important role in the detoxification and accumulation of the metals in plants (Piechalak et al. 2002). PCs can constitute metal-PC complexes and transport Pb to vacuoles and chloroplasts (Piechalak et al. 2002; Estrella-Gómez et al. 2009). The biosynthesis of PCs is catalyzed by phytochelatin synthase enzyme (PCS) that connect of γ-Glu-Cys to substrate glutathione molecules (Estrella-Gómez et al. 2009; Fischer et al. 2014). Thus, PCs biosynthesis depends on glutathione level which can correlate with phytochelatin content and metal sequestration (Thangavel et al. 2007). Estrella-Gómez et al. (2009) revealed that excessive lead accumulation increased both phytochelatin synthase (PCS) activity and phytochelatins content in the root of *Salvinia minima* as known the Pb^{2+} hyperaccumulator aquatic fern.

4 Concluding Remarks

Lead is a non-essential nutrient for plants, but it is easily taken up by the plant roots and transported to aboveground parts at its very small amount. Because the majority of Pb accumulates in the roots, Pb mainly affects the plants through root systems. In this review, the effects of Pb stress on growth and different enzymes in the plant are explained, and the following points are illuminated based on the studies so far: (1) Pb stress inhibits plant growth through directly or indirectly by various mechanisms. Lead-induced growth inhibition depends on directly the alterations in cell division and elongation by decreasing cell wall plasticity or by inhibiting microtubule development. In addition, lead can inhibit growth indirectly by various mechanisms such as disturbing of mineral nutrition, disrupting of photosynthesis, reduction of water potential. (2) Lead might inhibit or induce the activity of key enzymes in physiological and biochemical processes such as antioxidative mechanism, photosynthesis, respiration, seed germination, protein synthesis, nitrogen assimilation, proline synthesis, and phytochelatin synthesis. However, the effects of Pb on all these processes in the plant might show alteration depending to concentration, duration of exposure, plant species and different plant organs.

References

Abdelkrim S, Jebara SH, Saadani O, Jebara M (2018) Potential of efficient and resistant plant growth-promoting rhizobacteria in lead uptake and plant defence stimulation in *Lathyrus sativus* under lead stress. Plant Biol 20:857–869

Agency for Toxic Substances and Disease Registry (ATSDR) (2017) Priority list of hazardous substances (Online: https://www.atsdr.cdc.gov/spl/#2017spl)

Alamri SA, Siddiqui MH, Al-Khaishany MY, Nasir Khan M, Ali HM, Alaraidh IA, Mateen M (2018) Ascorbic acid improves the tolerance of wheat plants to lead toxicity. J Plant Interact 13:409–419

Ali N, Masood S, Mukhtar T, Kamran MA, Rafique M, Munis MFH, Chaudhary HJ (2015) Differential effects of cadmium and chromium on growth, photosynthetic activity, and metal uptake of *Linum usitatissimum* in association with Glomus intraradices. Environ Monit Assess 187:311

Aliu S, Gashi B, Rusinovci I, Fetahu S, Vataj R (2013) Effects of some heavy metals in some morpho-physiological parameters in maize seedlings. Am J Biochem Biotechnol 9:27

Arshad M, Ali S, Noman A, Ali Q, Rizwan M, Farid M, Irshad MK (2016) Phosphorus amendment decreased cadmium (Cd) uptake and ameliorates chlorophyll contents, gas exchange attributes, antioxidants, and mineral nutrients in wheat (*Triticum aestivum* L) under Cd stress. Arch Agron Soil Sci 62:533–546

Ashraf U, Kanu AS, Mo ZW, Hussain S, Anjum SA, Khan I, Abbas RN, Tang X (2015) Lead toxicity in rice; effects, mechanisms and mitigation strategies-a mini review. Environ Sci Pollut Res 22:18318–18332

Atici O, Aar G, Battal P (2005) Changes in phytohormone contents in chickpea seeds germinating under lead or zinc stress. Biol Plant 49:215–222

Bai XY, Dong YJ, Wang QH, Xu LL, Kong J, Liu S (2015) Effects of lead and nitric oxide on photosynthesis, antioxidative ability, and mineral element content of perennial ryegrass. Biol Plant 59:163–170

Bali S, Jamwal VL, Kaur P, Kohli SK, Ohri P, Gandhi SG, Ahmad P (2019) Role of P-type ATPase metal transporters and plant immunity induced by jasmonic acid against Lead (Pb) toxicity in tomato. Ecotoxicol Environ Saf 174:283–294

Barceló JUAN, Poschenrieder C (1990) Plant water relations as affected by heavy metal stress. J Plant Nutr 13:1–37

Bazzaz FA, Carlson RW, Rolfe GL (1975) Inhibition of corn and sunflower photosynthesis to lead concentration. J Environ Qual 34:156–1588

Bharwana SA, Ali S, Farooq MA, Iqbal N, Abbas F, Ahmad MSA (2013) Alleviation of lead toxicity by silicon is related to elevated photosynthesis, antioxidant enzymes suppressed lead uptake and oxidative stress in cotton. J Bioremed Biodegr 4:187

Brunet J, Repellin A, Varrault G, Terryn N, Zuily-Fodil Y (2008) Lead accumulation in the roots of grass pea (*Lathyrus sativus* L.): a novel plant for phytoremediation systems? C R Biol 331:859–864

Brunet J, Varrault G, Zuily-Fodil Y, Repellin A (2009) Accumulation of lead in the roots of grass pea (*Lathyrus sativus* L.) plants triggers systemic variation in gene expression in the shoots. Chemosphere 77:1113–1120

Burzynski M (1987) Influence of lead and cadmium on the absorption and distribution of potassium, calcium, magnesium and iron in cucumber seedlings. Acta Physiol Plant 9:229–238

Burzyński M, Grabowski A (1984) Influence of lead on N03 uptake and reduction in cucumber seedlings. Acta Soc Bot Pol 53:77–86

Cenkci S, Ciğerci İH, Yıldız M, Özay C, Bozdağ A, Terzi H (2010) Lead contamination reduces chlorophyll biosynthesis and genomic template stability in *Brassica rapa* L. Environ Exp Bot 67:467–473

Chatterjee C, Dube BK, Sinha P, Srivastava P (2004) Detrimental effects of lead phytotoxicity on growth, yield, and metabolism of rice. Commun Soil Sci Plant Anal 35:255–265

Checchi PM, Nettles JH, Zhou J, Snyder JP, Joshi HC (2003) Microtubule-interacting drugs for cancer treatment. Trends Pharmacol Sci 24:361–365

Chen Q, Zhang X, Liu Y, Wei J, Shen W, Shen Z, Cui J (2017) Hemin-mediated alleviation of zinc, lead and chromium toxicity is associated with elevated photosynthesis, antioxidative capacity; suppressed metal uptake and oxidative stress in rice seedlings. Plant Growth Regul 81:253–264

Chen Z, Yang B, Hao Z, Zhu J, Zhang Y, Xu T (2018) Exogenous hydrogen sulfide ameliorates seed germination and seedling growth of cauliflower under lead stress and its antioxidant role. J Plant Growth Regul 37:5–15

Choudhury S, Panda SK (2005) Toxic effects, oxidative stress and ultrastructural changes in moss *Taxithelium nepalense* (Schwaegr) Broth under chromium and lead phytotoxicity. Water Air Soil Pollut 167:73–90

Dalyan E, Yüzbaşıoğlu E, Akpınar I (2018) Effect of 24-Epibrassinolide on antioxidative defence system against lead-induced oxidative stress in the roots of *Brassica juncea* L. seedlings. Russ J Plant Physiol 65:570–578

Drazkiewicz M (1994) Chlorophyllase: occurrence, functions, mechanism of action, effects of external and internal factors. Photosynthetica 30:321–331

Đurđević B, Lisjak M, Stošić M, Engler M, Popović B (2008) Influence of Pb And Cu toxicity on lettuce photosynthetic pigments and dry matter accumulation. Cereal Res Commun 36:1951–1954

El-Beltagi HS, Mohamed AA (2010) Changes in non protein thiols, some antioxidant enzymes activity and ultrastructural alteration in radish plant (*Raphanus sativus* L) grown under lead toxicity. Not Bot Horti Agrobo 38:76–85

Estrella-Gómez N, Mendoza-Cózatl D, Moreno-Sánchez R, González-Mendoza D, Zapata-Pérez O, Martínez-Hernández A, Santamaría JM (2009) The Pb-hyperaccumulator aquatic fern Salvinia minima Baker, responds to Pb^{2+} by increasing phytochelatins via changes in SmPCS expression and in phytochelatin synthase activity. Aquat Toxicol 91:320–328

Fischer S, Kühnlenz T, Thieme M, Schmidt H, Clemens S (2014) Analysis of plant Pb tolerance at realistic submicromolar concentrations demonstrates the role of phytochelatin synthesis for Pb detoxification. Environ Sci Technol 48:7552–7559

Ghani A, Shah AU, Akhtar U (2010) Effect of lead toxicity on growth, chlorophyll and lead (Pb). Pak J Nutr 9:887–891

Gichner T, Žnidar I, Száková J (2008) Evaluation of DNA damage and mutagenicity induced by lead in tobacco plants. Mutat Res Genet Toxicol Environ 652:186–190

Gill SS, Tuteja N (2010) Reactive oxygen species and antioxidant machinery in abiotic stress tolerance in crop plants. Plant Physiol Biochem 48:909–930

Gopal R, Rizvi AH (2008) Excess lead alters growth, metabolism and translocation of certain nutrients in radish. Chemosphere 70:1539–1544

Gupta DK, Nicoloso FT, Schetinger MRC, Rossato LV, Pereira LB, Castro GY, Tripathi RD (2009) Antioxidant defense mechanism in hydroponically grown *Zea mays* seedlings under moderate lead stress. J Hazard Mater 172:479–484

Gupta DK, Huang HG, Yang XE, Razafindrabe BHN, Inouhe M (2010) The detoxification of lead in *Sedum alfredii* H is not related to phytochelatins but the glutathione. J Hazard Mater 177:437–444

Hadi F, Aziz T (2015) A mini review on lead (Pb) toxicity in plants. J Biol Life Sci 6:91–101

Hao F, Wang X, Chen J (2006) Involvement of plasma-membrane NADPH oxidase in nickel-induced oxidative stress in roots of wheat seedlings. Plant Sci 170:151–158

Harpaz-Saad S, Azoulay T, Arazi T, Ben-Yaakov E, Mett A, Shiboleth YM, Eyal Y (2007) Chlorophyllase is a rate-limiting enzyme in chlorophyll catabolism and is posttranslationally regulated. Plant Cell 19:007–1022

Hasanuzzaman M, Hossain MA, Teixeira da Silva JA, Fujita M (2012) Plant response and tolerance to abiotic oxidative stress: antioxidant defense is a key factor. In: Venkateswarlu B, Shanker A, Shanker C, Maheswari M (eds) Crop stress and its management: perspectives and strategies. Springer, Dordrecht, pp 261–315

Hasanuzzaman M, Nahar K, Rahman A, Mahmud JA, Alharby HF, Fujita M (2018) Exogenous glutathione attenuates lead-induced oxidative stress in wheat by improving antioxidant defense and physiological mechanisms. J Plant Interac 13:203–212

Hattab S, Flores-Casseres ML, Boussetta H, Doumas P, Hernandez LE, Banni M (2016) Characterisation of lead-induced stress molecular biomarkers in *Medicago sativa* plants. Environ Exp Bot 123:1–12

Hou X, Han H, Cai L, Liu A, Ma X, Zhou C, Meng F (2018) Pb stress effects on leaf chlorophyll fluorescence, antioxidative enzyme activities, and organic acid contents of *Pogonatherum crinitum* seedlings. Flora 240:82–88

Huang H, Gupta DK, Tian S, Yang XE, Li T (2012) Lead tolerance and physiological adaptation mechanism in roots of accumulating and non-accumulating ecotypes of *Sedum alfredii*. Environ Sci Pollut Res 19:1640–1651

Hussain A, Abbas N, Arshad F, Akram M, Khan ZI, Ahmad K, Mirzaei F (2013) Effects of diverse doses of Lead (Pb) on different growth attributes of *Zea-Mays* L. Agric Sci 4:262–265

Islam E, Liu D, Li T, Yang X, Jin X, Mahmood Q, Tian S, Li J (2008) Effect of Pb toxicity on leaf growth, physiology and ultrastructure in the two ecotypes of *Elsholtzia argyi*. J Hazard Mater 154:914–926

Jayasri MA, Suthindhiran K (2016) Effect of zinc and lead on the physiological and biochemical properties of aquatic plant *Lemna minor*: its potential role in phytoremediation. Appl Water Sci 7:1247–1253

Jiang Z, Qin R, Zhang H, Zou J, Shi Q, Wang J, Liu D (2014) Determination of Pb genotoxic effects in Allium cepa root cells by fluorescent probe, microtubular immunofluorescence and comet assay. Plant and Soil 383:357–372

John R, Ahmad P, Gadgil K, Sharma S (2009) Cadmium and lead-induced changes in lipid peroxidation, antioxidative enzymes and metal accumulation in *Brassica juncea* L. at three different growth stages. Arch Agron Soil Sci 55:395–405

Kaur G, Singh HP, Batish DR, Kumar Kohli R (2012a) Growth, photosynthetic activity and oxidative stress in wheat (*Triticum aestivum*) after exposure of lead to soil. J Environ Biol 33:265–269

Kaur G, Singh HP, Batish DR, Kohli RK (2012b) A time course assessment of changes in reactive oxygen species generation and antioxidant defense in hydroponically grown wheat in response to lead ions (Pb $^{2+}$). Protoplasma 249:1091–1100

Kaur G, Singh HP, Batish DR, Kohli RK (2014) Pb-inhibited mitotic activity in onion roots involves DNA damage and disruption of oxidative metabolism. Ecotoxicology 23:1292–1304

Khan M, Daud MK, Basharat A, Khan MJ, Azizullah A, Muhammad N (2016) Alleviation of lead-induced physiological, metabolic, and ultramorphological changes in leaves of upland cotton through glutathione. Environ Sci Pollut Res 23:8431–8440

Khan MN, Mobin M, Abbas ZK, Siddiqui MH (2017) Nitric oxide-induced synthesis of hydrogen sulfide alleviates osmotic stress in wheat seedlings through sustaining antioxidant enzymes, osmolyte accumulation and cysteine homeostasis. Nitric Oxide 68:91–102

Khan F, Hussain S, Tanveer M, Khan S, Hussain HA, Iqbal B, Geng M (2018a) Coordinated effects of lead toxicity and nutrient deprivation on growth, oxidative status, and elemental composition of primed and non-primed rice seedlings. Environ Sci Pollut Res 25:21185–21194

Khan MM, Islam E, Irem S, Akhtar K, Ashraf MY, Iqbal J, Liu D (2018b) Pb-induced phytotoxicity in para grass (*Brachiaria mutica*) and Castorbean (*Ricinus communis L*) antioxidant and ultrastructural studies. Chemosphere 200:257–265

Kohli SK, Handa N, Sharma A, Gautam V, Arora S, Bhardwaj R, Wijaya L, Alyemeni MN, Ahmad P (2018) Interaction of 24-epibrassinolide and salicylic acid regulates pigment contents, antioxidative defense responses, and gene expression in *Brassica juncea* L. seedlings under Pb stress. Environ Sci Pollut Res 25:15159–15173

Kozhevnikova AD, Seregin IV, Bystrova EI, Belyaeva AI, Kataeva MN, Ivanov VB (2009) The effects of lead, nickel, and strontium nitrates on cell division and elongation in maize roots. Russ J Plant Physiol 56:242–250

Kumar A, Majeti NVP (2014) Proteomic responses to lead-induced oxidative stress in *Talinum triangulare* Jacq (Willd) roots: identification of key biomarkers related to glutathione metabolisms. Environ Sci Pollut Res 21:8750–8764

Kumar A, Prasad MNV (2018) Plant–lead interactions: transport, toxicity, tolerance, and detoxification mechanisms. Ecotoxicol Environ Saf 166:401–418

Kumar A, Pal L, Agrawal V (2017) Glutathione and citric acid modulates lead- and arsenic-induced phytotoxicity and genotoxicity responses in two cultivars of *Solanum lycopersicum* L. Acta Physiol Planta 39:151

Lamhamdi M, Bakrim A, Aarab A, Lafont R, Sayah F (2011) Lead phytotoxicity on wheat (*Triticum aestivum* L) seed germination and seedlings growth. C R Biol 334:118–126

Lamhamdi M, El Galiou O, Bakrim A, Nóvoa-Muñoz JC, Arias-Estevez M, Aarab A, Lafont R (2013) Effect of lead stress on mineral content and growth of wheat (*Triticum aestivum*) and spinach (*Spinacia oleracea*) seedlings. Saudi J Biol Sci 20:29–36

Liu D, Li TQ, Jin XF, Yang XE, Islam E, Mahmood Q (2008) Lead induced changes in the growth and antioxidant metabolism of the lead accumulating and non-accumulating ecotypes of *Sedum alfredii*. J Integr Plant Biol 50:129–140

Lopez ML, Peralta-Videa JR, Benitez T, Duarte-Gardea M, Gardea-Torresdey JL (2007) Effects of lead, EDTA, and IAA on nutrient uptake by alfalfa plants. J Plant Nutr 30:1247–1261

Maier R (1978) Studies on the effect of lead in the acid phosphatase in *Zea mays* L. Z Pflanzenphysiol 87:347–354

Malar S, Manikandan R, Favas PJ, Sahi SV, Venkatachalam P (2014a) Effect of lead on phytotoxicity, growth, biochemical alterations and its role on genomic template stability in *Sesbania grandiflora*: a potential plant for phytoremediation. Ecotoxicol Environ Saf 108:249–257

Malar S, Vikram SS, Favas PJC, Perumal V (2014b) Lead heavy metal toxicity induced changes on growth and antioxidative enzymes level in water hyacinths [*Eichhornia crassipes* (Mart.)]. Bot Stud 55:54

Małecka A, Derba-Maceluch M, Kaczorowska K, Piechalak A, Tomaszewska B (2009) Reactive oxygen species production and antioxidative defense system in pea root tissues treated with lead ions: mitochondrial and peroxisomal level. Physiol Plant 31:1065–1075

Mika A, Minibayeva F, Beckett R, Lüthje S (2004) Possible functions of extracellular peroxidases in stress-induced generation and detoxification of active oxygen species. Phytochem Rev 3:173–193

Mishra S, Srivastava S, Tripathi RD, Kumar R, Seth CS, Gupta DK (2006) Lead detoxification by coontail (*Ceratophyllum demersum* L) involves induction of phytochelatins and antioxidant system in response to its accumulation. Chemosphere 65:1027–1039

Mroczek-Zdyrska M, Strubińska J, Hanaka A (2017) Selenium improves physiological parameters and alleviates oxidative stress in shoots of lead-exposed *Vicia faba* L minor plants grown under phosphorus-deficient conditions. J Plant Growth Regul 36:186–199

Mukherji S, Maitra P (1976) Toxic effects of lead on growth metabolism of germinating rice (*Oryza sativa* L) seeds on mitosis of onion (*Allium cepa* L) root tip cells. Indian J Exp Biol 14:519–521

Nareshkumar A, Veeranagamallaiah G, Pandurangaiah M, Kiranmai K, Amaranathareddy V, Lokesh U, Sudhakar C (2015) Pb-stress induced oxidative stress caused alterations in antioxidant efficacy in two groundnut (*Arachis hypogaea* L) cultivars. Agric Sci 6:1283–1297

Olmos E, Martínez-Solano JR, Piqueras A, Hellín E (2003) Early steps in the oxidative burst induced by cadmium in cultured tobacco cells (BY-2 line). J Exp Bot 54:291–300

Päivöke A (1983) The long-term effects of lead and arsenate on the growth and development, chlorophyll content and nitrogen fixation of the garden pea. Ann Bot Fennici 20:297–306

Päivöke AE (2002) Soil lead alters phytase activity and mineral nutrient balance of *Pisum sativum*. Environ Exp Bot 48:61–73

Pandey S, Gupta K, Mukherjee AK (2007) Impact of cadmium and lead on *Catharanthus roseus*-a phytoremediation study. J Environ Biol 28:655–662

Parys E, Wasilewska W, Siedlecka M, Zienkiewicz M, Drożak A, Romanowska E (2014) Metabolic responses to lead of metallicolous and nonmetallicolous populations of *Armeria maritima*. Arch Environ Contam Toxicol 67:565–577

Piechalak A, Tomaszewska B, Baralkiewicz D, Malecka A (2002) Accumulation and detoxification of lead ions in legumes. Phytochemistry 60:153–162

Piotrowska A, Bajguz A, Godlewska-Żyłkiewicz B, Czerpak R, Kamińska M (2009) Jasmonic acid as modulator of lead toxicity in aquatic plant *Wolffia arrhiza* (Lemnaceae). Environ Exp Bot 66:507–513

Porter JR, Sheridan RP (1981) Inhibition of nitrogen fixation in alfalfa by arsenate, heavy metals, fluoride, and simulated acid rain. Plant Physiol 68:143–148

Potters G, Horemans N, Jansen MA (2010) The cellular redox state in plant stress biology–a charging concept. Plant Physiol Biochem 48:292–300

Pourrut B (2008) Implication du stress oxydatif dans la toxicité du plomb sur une plante modèle, *Vicia faba*. Doctoral dissertation, Institut National Polytechnique de Toulouse, France

Pourrut B, Jean S, Silvestre J, Pinelli E (2011) Lead-induced DNA damage in Vicia faba root cells: potential involvement of oxidative stress. Mutat Res Genet Toxicol Environ 726:123–128

Quartacci MF, Cosi E, Navari-Izzo F (2001) Lipids and NADPH-dependent superoxide production in plasma membrane vesicles from roots of wheat grown under copper deficiency or excess. J Exp Bot 52:77–84

Qureshi MI, Abdin MZ, Qadir S, Iqbal M (2007) Lead-induced oxidative stress and metabolic alterations in *Cassia angustifolia* Vahl. Biol Plant 51:121–128

Reddy AM, Kumar SG, Jyothsnakumari G, Thimmanaik S, Sudhakar C (2005) Lead induced changes in antioxidant metabolism of horsegram (*Macrotyloma uniflorum* (Lam) Verdc) and bengalgram (*Cicer arietinum* L). Chemosphere 60:97–104

Riffat J, Ahmad P, Gadgil K, Sharma S (2009) Cadmium and lead-induced changes in lipid peroxidation, antioxidative enzymes and metal accumulation in *Brassica juncea* L. at three different growth stages. Arch Agron Soil Sci 55:395–405

Rizwan M, Ali S, Rehman MZ, Javed MR, Bashir A (2018) Lead toxicity in cereals and its management strategies: a critical review. Water Air Soil Pollut 229:211

Rodriguez E, Azevedo R, Moreira H, Sout L, Santos C (2013) Pb^{2+} exposure induced microsatellite instability in *Pisum sativum* in a locus related with glutamine metabolism. Plant Physiol Biochem 62:19–22

Romanowska E, Igamberdiev AU, Parys E, Gardeström P (2002) Stimulation of respiration by Pb^{2+} in detached leaves and mitochondria of C3 and C4 plants. Physiol Plant 116:148–154

Rucińska R, Waplak S, Gwóźdź EA (1999) Free radical formation and activity of antioxidant enzymes in lupin roots exposed to lead. Plant Physiol Biochem 37:187–194

Rucińska R, Sobkowiak R, Gwóźdź EA (2004) Genotoxicity of lead in lupin root cells as evaluated by the comet assay. Cell Mol Biol Lett 9:519–528

Rucińska-Sobkowiak R, Nowaczyk G, Krzesłowska M, Rabęda I, Jurga S (2013) Water status and water diffusion transport in lupine roots exposed to lead. Environ Exp Bot 87:100–109

Ruley T, Sharma NC, Sahi SV (2004) Antioxidant defense in a lead accumulating plant, *Sesbania drummondii*. Plant Physiol Biochem 42:899–906

Sengar RS, Gautam M, Sengar RS, Sengar RS, Garg SK, Sengar K, Chaudhary R (2009) Lead stress effects on physiobiochemical activities of higher plants. Rev Environ Contam Toxicol 196:1–21

Seregin IV, Ivanov VB (2001) Physiological aspects of cadmium and lead toxic effects on higher plants. Russ J Plant Physiol 48:523–544

Seregin IV, Kozhevnikova AD (2006) Physiological role of nickel and its toxic effects on higher plants. Russ J Plant Physiol 53:257–277

Shahid M, Pinelli E, Pourrut B, Silvestre J, Dumat C (2011) Lead-induced genotoxicity to *Vicia faba* L roots in relation with metal cell uptake and initial speciation. Ecotoxicol Environ Saf 74:78–84

Shakoor MB, Ali S, Hameed A, Farid M, Hussain S, Yasmeen T, Abbasi GH (2014) Citric acid improves lead (Pb) phytoextraction in *Brassica napus* L by mitigating Pb-induced morphological and biochemical damages. Ecotoxicol Environ Saf 109:38–47

Sharma P, Dubey RS (2005) Lead toxicity in plants. Braz J Plant Physiol 17:35–52

Shu X, Yin L, Zhang Q, Wang W (2012) Effect of Pb toxicity on leaf growth, antioxidant enzyme activities, and photosynthesis in cuttings and seedlings of *Jatropha curcas* L. Environ Sci Pollut Res 19:893–902

Sidhu GPS, Singh HP, Batish DR, Kohli RK (2016) Effect of lead on oxidative status, antioxidative response and metal accumulation in *Coronopus didymus*. Plant Physiol Biochem 105:290–296

Silva S, Silva P, Oliveira H, Gaivão I, Matos M, Pinto-Carnide O, Santos C (2017) Pb low doses induced genotoxicity in *Lactuca sativa* plants. Plant Physiol Biochem 112:109–116

Singh R, Tripathi RD, Dwivedi S, Kumar A, Trivedi PK, Chakrabarty D (2010) Lead bioaccumulation potential of an aquatic macrophyte *Najas indica* are related to antioxidant system. Bioresour Technol 101:3025–3032

Singh S, Srivastava PK, Kumar D, Tripathi DK, Chauhan DK, Prasad SM (2015) Morpho-anatomical and biochemical adapting strategies of maize (*Zea mays* L) seedlings against lead and chromium stresses. Biocatal Agric Biotechnol 4:286–295

Stefanov K, Seizova K, Popova I, Petkov V, Kimenov G, Popov S (1995) Effect of lead ions on the phospholipid composition in leaves of *Zea mays* and *Phaseolus vulgaris*. J Plant Physiol 147:243–246

Sudhakar R, Venu G (2001) Mitotic abnormalities induced by silk dyeing industry effluents in the cells of *Allium cepa*. Cytologia 66:235–239

Tariq SR, Rashid N (2013) Multivariate analysis of metal levels in paddy soil, rice plants, and rice grains: a case study from Shakargarh, Pakistan. J Chem 2013:539251

Thakur S, Singh L, Zularisam AW, Sakinah M, Din MFM (2017) Lead induced oxidative stress and alteration in the activities of antioxidative enzymes in rice shoots. Biol Plant 61:595–598

Thangavel P, Long S, Minocha R (2007) Changes in phytochelatins and their biosynthetic inter-mediates in red spruce (*Picea rubens Sarg*) cell suspension cultures under cadmium and zinc stress. Plant Cell Tiss Org Cult 88:201–216

Tu SI, Brouillette JN (1987) Metal ion inhibition of corn root plasma membrane ATPase. Phytochemistry 26:65–69

Türkoğlu S (2012) Determination of genotoxic effects of chlorfenvinphos and fenbuconazole in *Allium cepa* root cells by mitotic activity, chromosome aberration, DNA content, and comet assay. Pestic Biochem Physiol 103:224–230

Venkatachalam P, Jayalakshmi N, Geetha N, Sahi SV, Sharma NC, Rene ER, Favas PJ (2017) Accumulation efficiency, genotoxicity and antioxidant defense mechanisms in medicinal plant *Acalypha indica L* under lead stress. Chemosphere 171:544–553

Verma S, Dubey RS (2003) Lead toxicity induces lipid peroxidation and alters the activities of antioxidant enzymes in growing rice plants. Plant Sci 164:645–655

Wang CR, Wang XR, Tian Y, Yu HX, Gu XY, Du WC, Zhou H (2008) Oxidative stress, defense response, and early biomarkers for lead-contaminated soil in *Vicia faba* seedlings. Environ Toxicol Chem 27:970–977

Weryszko-Chmielewska E, Chwil M (2005) Lead-induced histological and ultrastructural changes in the leaves of soybean (*Glycine max* (L.) Merr.). Soil Sci Plant Nutr 51:203–212

Wierzbicka M (1998) Lead in the apoplast of *Allium cepa* L root tips—ultrastructural studies. Plant Sci 133:105–119

Xiong ZT, Zhao F, Li MJ (2006) Lead toxicity in *Brassica* pekinensis Rupr: effect on nitrate assimilation and growth. Environ Toxicol 21:147–153

Yandow TS, Klein RM (1986) Nitrate reductase of primary roots of red spruce seedlings: effects of acidity and metal ions. Plant Physiol 81:723–725

Yang Y, Zhang Y, Wei X, You J, Wang W, Lu J, Shi R (2011) Comparative antioxidative responses and proline metabolism in two wheat cultivars under short term lead stress. Ecotoxicol Environ Saf 74:733–740

Zhang Y, Deng B, Li Z (2018) Inhibition of NADPH oxidase increases defense enzyme activities and improves maize seed germination under Pb stress. Ecotoxicol Environ Saf 158:187–192

Zhou C, Huang M, Li Y, Luo J, Ping Cai L (2016) Changes in subcellular distribution and anti-oxidant compounds involved in Pb accumulation and detoxification in *Neyraudia reynaudiana*. Environ Sci Pollut Res 23:21,794–21,804

Biological Strategies of Lichen Symbionts to the Toxicity of Lead (Pb)

Joana R. Expósito, Eva Barreno, and Myriam Catalá

Abstract Lichens are symbiotic organisms, originated by mutualistic associations of heterotrophic fungi (mycobiont), photosynthetic partners (photobionts) which can be either cyanobacteria (cyanobionts) or green microalgae (phycobionts), and bacterial consortia. They are poikilohydric organisms without cuticles or nutrient absorption organs adapted to anhydrobiosis. They present a large range of tolerance to abiotic stress (UV radiation, extreme temperatures, high salinity, mineral excess, etc.) and prosper all around the Earth, especially in harsh habitats, including Antarctica and warm deserts. Their biodiversity is widely used as a bioindicator of environmental quality due to this diversity of tolerance in different species, and they are included in air Pb monitoring programmes worldwide. Their ability to bioaccumulate environmental substances, including some air pollutants and heavy metals, makes them excellent passive biomonitors of Pb. Heavy metal tolerance is related to diverse mechanisms: cell walls and exclusion systems (such as extracellular polymeric substances), intracellular chelators and an extraordinary antioxidant and repair capacity. But recent data show that the most powerful mechanism is related with the upregulation of mutual systems by symbiosis.

Keywords Pb · Lichens · Bioindicators · Biomonitors · Tolerance · Microalgae

J. R. Expósito · M. Catalá (✉)
Departamento de Biología y Geología, Física y Química Inorgánica, (ESCET),
Universidad Rey Juan Carlos, Móstoles, Madrid, Spain
e-mail: myriam.catala@urjc.es

E. Barreno
Universitat de València, ICBIBE (Instituto Cavanilles de Biodiversidad y Biología Evolutiva)
Botánica, Facultad de C. Biológicas, Burjassot, Valencia, Spain

© Springer Nature Switzerland AG 2020
D. K. Gupta et al. (eds.), *Lead in Plants and the Environment*, Radionuclides
and Heavy Metals in the Environment,
https://doi.org/10.1007/978-3-030-21638-2_9

149

1   Introduction

1.1   Definition, Biological and Ecological Characteristics of Lichens

Lichens originate from mutualistic symbiosis, involving at least a fungus "mycobiont" (heterotrophic, the dominant partner) and one or several photosynthetic "photobionts" (the carbohydrate producers), either green microalgae (phycobionts) or cyanobacteria (cyanobionts) or both. Also, specific bacterial communities are obligate lichen symbionts and, therefore, considered to be an integral part of lichen thalli (García-Breijo et al. 2010; Aschenbrenner et al. 2016). These symbioses are also cyclical as the three partners must activate the association with every new generation (Margulis and Barreno 2003). In thallus (holobiont), the mycobiont supplies water, mineral substances and protection against adverse factors; while the photobionts and the bacteria provide organic and inorganic nutrients (Nash III 1996; Honegger 1998; Cernava et al. 2017).

Lichens are unique symbiotic phenotypes with specific structure (holotypes) and complex physiology which is based on the functional and genetic interactions between symbionts (Barreno 2013), and are evaluated as microecosystems (Honegger 1991; Honegger 1998). Like whole microecosystems, lichens cannot be classified in natural systems as species can (Tehler and Irestedt 2007). However, lichens are usually known as lichenized fungi and are classified according to the type of mycobiont. In this way, the classification of lichens is included within the systematics of fungi. Lichens are polyphyletic and each species corresponds to a different mycobiont. Three basic lichen growth forms (architectural types) are known: crustose, foliose and fruticose with numerous transitions. These three-dimensional forms have been used to classify lichens at family or genus level (Tehler and Irestedt 2007). There are some exceptions, for example, *Cladonia* and *Stereocaulon* (Högnabba 2006) are dimorphic with basal grains or squamules and fruticose structures.

1.2   Lichens and Environment

The lichenization permits the symbionts to grow in extreme environments turning lichens into pioneers of fresh rock outcrops. They dominate 6% of the land's surface, and contribute to soil formation (Purvis and Pawlik-Skowrońska 2008). Due to these facts, lichens prosper all around the Earth and in the majority of habitats, including Antarctica and deserts. This development would not be possible for each symbiont on their own (Ahmadjian 1995). Altitude, contamination and anthropogenic pressure are key to understand the diversity and distributions of these organisms (Firdous et al. 2017). From 13,500 species to about 20,000, if orphan or hardly recorded species are included, have been estimated (Sipman and Aptroot 2001; Kirk et al. 2008; Lücking et al. 2016).

1.2.1 Tolerance of Lichens to Abiotic Stress

Either morphological, physiological or ecological features are necessary for lichens to tolerate extreme conditions and thus be adapted to an extreme environment (Margulis and Barreno 2003). For example, lichens surviving in humid habitats have to be adapted to shade (thinner thalli and less pigmented), have a waterproof surface and endure elevated temperatures while metabolically active. To tolerate Alpine regions, lichens take advantage of short favourable periods, amass phenolic substances as usnic acid that absorbs UV and change their composition of proteins and fatty acids with altitude. Those that tolerate desert conditions have thin crustose growth forms, low water potentials and, sometimes, photoprotective pigments. Lichens are also frequently found in chemically rich environments (high salinity, pollution or heavy metals presence). In order to survive they need to tolerate high or low pH, have antioxidant resistance mechanisms and a "breathable" and "rough" surface, among other adaptations (Armstrong 2017).

The main physiological trait of lichens is that, unlike plants that are isolated by waxy cuticles, they are poikilohydrous, which means that their water content is in equilibrium with the environment. Therefore, they are subjected to continuous desiccation/rehydration cycles balancing their hydric potential with that of the atmosphere. They are adapted to anhydrobiosis and can resume cell function immediately after rehydrating (Green et al. 2011).

As stated above, lichens inhabit extremely harsh areas, but they can metabolize and photosynthesize under harsh conditions, such as low temperatures, aridity and high UV radiation flows. High concentrations of salt affect poikilohydric organisms, making them lose intracellular water (Hasegawa and Bressan 2000). In intertidal zones, salinity is the key to determining the distribution of lichens (Delmail et al. 2013). Even, some lichen species can survive simulated and real outer space conditions (de Vera 2012; de Vera et al. 2019).

Tolerance to desiccation in lichens has only now been addressed by researchers and is still being revealed. Solar radiation and desiccation increase ROS that can result in oxidative stress (Kranner et al. 2005). Therefore it must be linked to some kind of mechanism based on reparation after rehydration, since chlorophyll fluorescence recovers almost immediately (Calatayud et al. 1997; Beckett et al. 2005). Still, recent studies show that this tolerance is also due to protective mechanisms that eliminate reactive oxygen species (ROS) or prevent their formation such as the presence of antioxidants, antioxidant enzymes and NO, among others (Beckett et al. 2008; Kranner et al. 2008; Catalá et al. 2010b).

1.2.2 Tolerance to Heavy Metals

As mentioned above, in areas with high concentrations of metals we can find abundant lichens, including rare species restricted to these environments. On the other hand, lichens are exposed to anthropogenic environmental heavy metals as lead by the same routes than the rest of photosynthetic biota. Humid and dry deposition of aerosols emitted by metallurgical factories, mining (refinery and smelting operations), combustion (i.e. incineration, leaded gasoline, coal power plants, fires) and general urban and industrial activities are long recognized as pollution sources (reviewed in Bollhöfer and Rosman 2002; Zhang et al. 2019).

Mercury, cadmium or lead have no known biological function in lichens, nonetheless, due to their lack of protective structures they penetrate thalli and bioaccumulate within cells in a significantly more effective manner than in plants (Laaksovirta et al. 1976; Arhoun et al. 2000; Augusto et al. 2010). Heavy metal concentration inside lichen thalli is well correlated with the distance to the pollution source as well as with climatic and land cover factors, greatly impacting on lichen abundance and richness (Branquinho et al. 1999; Pinho et al. 2008). One of the key markers of anthropogenic pollution on the earth is the decrease of lichen diversity which was already shown in the late 19th century in Paris (Ellis and Coppins 2009; Agnan et al. 2017).

Despite lichens are bioaccumulators of trace elements and show a relative high resistance to heavy metals, not all lichens are tolerant to heavy metals and there is a vast variability in tolerance (Puckett 1976; Brown and Beckett 1983; Sarret et al. 1998; Giordani et al. 2002; Garty et al. 2003). Several authors have elaborated lists of lichen species classified attending to their relative tolerance to these toxicants (see Table 1) (Giordani et al. 2002; Loppi and Frati 2006; Agnan et al. 2017). In some lichens tolerance to metals is inducible, but in others, such as *Cladonia*, seems to be constitutive (Bačkor et al. 2011). Some species, such as *Lecanora polytropa* are even considered as hyperaccumulators (Purvis et al. 2008).

Knowing how lichens accumulate and endure toxic elements is necessary for biodiversity conservation and is also key to biomonitoring and even for biogeochemical cycles comprehension (Purvis and Pawlik-Skowrońska 2008). The form of growth seems to be important for natural heavy metal tolerance. Crustose lichens

Table 1 List of resistant, intermediate and sensitive lichen species relative to atmospheric pollution (Adapted from Agnan et al. 2017)

Resistant species	Intermediate species	Sensitive species
Acrocordia gemmata	*Amandinea punctata*	*Caloplaca ferruginea*
Arthonia radiata	*Buellia disciformis*	*Evernia prunastri*
Calicium salicinum	*Chrysothrix candelaris*	*Hypogymnia physodes*
Cladonia fimbriata	*Lecanora argentata*	*Pertusaria coccodes*
Dendrographa decolorans	*Lecanora barkmaniana*	*Physcia adscendens*
Graphis scripta	*Lecanora carpinea*	*Physconia distorta*
Lecanactis subabietina	*Lecanora chlarotera*	*Pleurosticta acetabulum*
Lecanora allophana	*Lecanora conizaeoides*	*Pseudevernia furfuracea*
Lepraria incana	*Lecanora expallens*	*Ramalina farinacea*
Melanohalea exasperatula	*Lecidella elaeochroma*	*Usnea sp.*
Ochrolechia androgyna	*Melanelixia glabratula*	
Ochrolechia pallescens subsp. parella	*Ochrolechia turneri*	
Pertusaria albescens	*Parmelia sulcata*	
Pertusaria amara	*Parmelina carporrhizans*	
Pertusaria leioplaca	*Phlyctis argena*	
Physcia tenella	*Xanthoria parietina*	
Schismatomma cretaceum		

have been proposed to be more tolerant than fruticose since they are more closely attached to the substrate and consequently more intensely exposed to toxic minerals (Beck 1999; Bačkor et al. 2010). Cellular heavy metal tolerance may be achieved by cytoplasmic immobilization, cytoplasmic tolerance or transport of ions to cell wall and the cell membrane (Armstrong 2017). However, there exists a debate about the mechanisms that each symbiont contributes for holobiont heavy metal tolerance (Álvarez et al. 2012, 2015). In general, phycobionts are usually more sensitive than mycobionts to metal contamination (Ahmadjian 1993). At least, they are more sensitive than the mycobionts to the presence of heavy metals when grown outside the symbiosis (Bačkor et al. 2006, 2007). Therefore, the ecological success of lichens living on substrates contaminated with toxic elements may be due to the presence of metal-tolerant phycobionts, however, contradicting data must still be considered (Bačkor et al. 2010; Álvarez et al. 2015).

1.3   Uses of Lichens in the Evaluation of Environmental Quality

Lichens are good indicators of global change because they are sensitive to variations in temperature, air pollution and water availability (Bajpai et al. 2018). Even subtle environmental changes affect measurably their biodiversity (Eldridge and Delgado-Baquerizo 2018). Urbanization and environmental pollution seem to affect all the functional features of lichens, with different responses according to the group (Koch et al. 2019). As a matter of fact, anthropogenic activities, and consequently, pollution and climate change, have been proposed to explain the structure of the lichen communities (McCune et al. 1997; Geiser and Neitlich 2007; Bajpai et al. 2018). Table 2 contains some examples of articles on lichens as bioindicators in different areas.

Lichens have long been used as environmental toxicity and air quality bioindicators given that, despite their outstanding stress tolerance, they are very sensitive to the accumulation of certain atmospheric pollutants (Hawksworth and Rose 1970; Herzig et al. 1990; Richardson 1993; Nash III and Gries 1995; Cislaghi and Nimis 1997; Cumming et al. 2007). On one hand, due to their physiology, they are bioaccumulators of trace elements (Garty et al. 1993; Sloof 1995; Loppi and Bargagli 1996; Bargagli 1998; Sarret et al. 1998), but they are also sensitive to toxic gases such as ozone, sulphur and nitrogen oxides (Conti and Cecchetti 2001; Giordani et al. 2002; Berganimini et al. 2005; Giordani et al. 2014; Ochoa-Hueso et al. 2017).

Lichens are ideal biomonitors, they occupy diverse regions allowing their comparison and they are submitted to contaminants throughout the year and their response is integrated along time (Markert 1993; Loppi et al. 2000). Passive biomonitoring with lichens is used worldwide as a biological measuring system for monitoring atmospheric pollution (Herzig et al. 1990; Loppi and Bargagli 1996; Giordani et al. 2002; Giordani 2007; Augusto et al. 2010; Agnan et al. 2017). Both the diversity of the lichen communities and the pollutants bioaccumulation may be used (Nimis et al. 1990; Sloof 1995; Komárek et al. 2008). Lichens are superb passive samplers of

Table 2 Lichens acting as bioindicators ecological parameters of lichen populations and communities used as surrogated measures of environmental quality

Community	Parameter	Region	References
Lichens	Abundance	Artic regions	Nash III and Gries (1995)
Dryland biocrust communities	Relative abundance	Australia	Eldridge and Delgado-Baquerizo (2018)
Lichens	Functional traits and functional groups	Brazil	Koch et al. (2019)
Lichen epiphytes associated with juniper scrub	Richness	Britain	Ellis and Coppins (2009)
Epiphytic lichens	Diversity	European forests	Stofer et al. (2012)
Lichenized fungi	Jaccard index	Floristic regions of the world (Takhtajan)	Feuerer and Hawksworth (2007)
Corticolous lichens	Diversity, Abundance and Shannon index	France and Switzerland	Agnan et al. (2017)
Epiphytic lichens	Diversity value, Relative diversity	Italy	Nimis et al. (1990), Giordani (2007)
Epiphytic lichens	Diversity	Mediterranean basin and woodlands	Pinho et al. (2011), Ochoa-Hueso et al. (2017)
Lichens	Diversity	Pakistan	Firdous et al. (2017)
Epiphytic lichens	Lichen Diversity value	Portugal	Pinho et al. (2008)
Lichens	Richness	Several European countries	Berganimini et al. (2005)
Epiphytic macrolichens	Richness	Southeast United States	McCune et al. (1997)
Epiphytic lichens	Communities on conifers	Southern California mountains	Sigal and Nash III (1983)
Gypsophilous lichens	Genetic diversity (Phylogenetic analysis)	Spain	Chiva et al. (2019)

atmosphere contamination both for inorganic and organic elements (Nash III and Gries 2002; Augusto et al. 2010) since their accumulation of xenobiotics is directly proportional to environmental concentrations (Ockenden et al. 1998; Poličnik et al. 2004; Sett and Kundu 2016). Table 3 presents some studies on lichens acting as biomonitors of heavy metals, air pollution and/or trace elements.

In particular, the biomonitoring of air heavy metals with lichens gathers abundant literature (Herzig et al. 1990; Loppi and Bargagli 1996; Loppi et al. 2000; Carreras and Pignata 2002; Poličnik et al. 2004; Kinalioğlu et al. 2006; Komárek et al. 2008; Bačkor and Loppi 2009; Garty 2010; Bosch-Roig et al. 2013; Sett and Kundu 2016). One outstanding case is the usage of these organisms as biomonitors of Pb as a marker for

Table 3 Lichens as passive biosensors. The table shows the organism that has been used as a biomonitor and the type of pollutant that accumulates heavy metals, air pollution and/or trace elements. The reference of each study is presented

Organism	Pollutant	References
Epiphytic lichens	Air particulate matter levels	Varela et al. (2018)
Epiphytic lichens	Nitrogen critical loads	Giordani et al. (2014)
Flavoparmelia caperata	Heavy metals	Loppi and Bargagli (1996), Loppi et al. (2000)
Flavoparmelia caperata	Trace elements	Paoli et al. (2012)
Hypogymnia physodes	Air pollution	Ockenden et al. (1998)
Hypogymnia physodes	Heavy metals	Herzig et al. (1990) Poličnik et al. (2004)
Parmelia sulcata	Trace elements	Sloof (1995)
Pseudevernia furfuracea	Heavy metals	Bari et al. (2001)
Punctelia subrudecta	Heavy metals	González and Pignata (1994)
Ramalina celastri	Heavy metals	Bérmudez et al. (2009)
Ramalina duriaei	Heavy metals	Garty and Fuchs (1982)
Ramalina fastigiata	Atmospheric metal	Branquinho et al. (1999)
Several species of lichens	Heavy metals	Agnan et al. (2015)
Several species of lichens	Heavy metals	Garty (2010)
Several species of lichens (review)	Air pollution Heavy metals	Conti and Cecchetti (2001)
Usnea amblyoclada	Air pollution Heavy metals	Carreras and Pignata (2001, 2002)
Xanthoria parietina	Trace elements	Nimis et al. (2000)

leaded gasoline. Lichens accumulate this metal correlating with traffic in the area (Takala and Olkkonen 1981; Garty et al. 1985). Studies carried out in 1970 showed an increase in Pb due to the massive use of tetraethyllead in gasoline, but the elimination of this octane booster in some countries caused a reduction in subsequent studies (Garty 2010). The diversity of lichens has even been recognized as a sensitive indicator of the biological effects of contaminants in human health (Cislaghi and Nimis 1997).

2 Lichens and Pb

2.1 *Damaging Effects of Pb on Lichens*

Despite lichens retain high concentrations of Pb extracellularly, mainly attached to cell walls and extracellular polymers, this metal has been shown to enter cells (Branquinho et al. 1997), especially in fungal cortical areas (Garty and Theiss 1990). Thanks to the big fungal biomass and morphology, lower quantities reach the phycobiont layer but even trace amounts of Pb cause alterations in lichen photosynthesis, because it affects chlorophyll integrity and content, especially at low pH (Garty et al. 1992; Garty et al. 1997; Branquinho et al. 1997). These effects on chlorophyll are related with a drop in PSII and photochemical reactions, particularly in cyanobionts. Namely, Branquinho et al. (1997) report that the important decrease in Fv/Fm observed for *Lobaria pulmonaria* may be related to a higher sensitivity to Pb of *Dictyochloropsis* microalgae with regard to *Trebouxia* (*Flavoparmelia caperata* and *Ramalina farinacea*). This higher sensitivity of prokaryotic photobionts to heavy metals has also been reported by other researchers (Puckett 1976; Brown and Beckett 1983; Garty et al. 1992; Branquinho and Brown 1994).

Some authors have related accumulated high concentrations of Pb among other metals with membrane damage leading to ion leaking or water loss (Puckett 1976; Chettri and Sawidis 1997; Garty et al. 1997; Chettri et al. 1998) but others have not (Branquinho et al. 1997). These divergences may be due to heavy metal dose, given that Puckett (Puckett 1976) observed that Pb initiated a dramatic K^+ efflux at a certain concentration, in contrast with other metals which induced a dose-dependent effect. Membrane damage may be mediated by lipid peroxidation, which is one of the main effects of prooxidant pollutants studied in lichens (Paoli et al. 2011; Gurbanov and Unal 2019). Whereas short time exposure (minutes to hours) to Pb has been shown to induce a compensatory hormetic decrease of lipid peroxidation in *Ramalina farinacea* (Álvarez et al. 2015), longer exposures (days) showed a positive correlation in *Xanthoria parietina* or *Cladonia convoluta* thalli (Dzubaj et al. 2008; Gurbanov and Unal 2019).

Infrared spectroscopic studies showed that lead exposure of *Cladonia convoluta* diminished the band corresponding to total nucleic acids. This decrease was especially related with the B-form of DNA, less stable under abiotic stress than the A-form, but involved in replication and transcription processes (Gurbanov and Unal 2019). The same study also reported an increase in the quantity of usnic acid, an important lichen substance involved in the extracellular heavy metal chelating ability of mycobionts (Purvis and Pawlik-Skowrońska 2008). In plants, high levels of Pb also cause inhibition of enzyme activities by reacting with their sulfhydryl groups, and disturb mineral nutrition (reviewed in Nagajyoti et al. 2010) but adequate studies addressing these effects in lichens must still be performed.

2.2 Exclusion Systems and Walls

Cell wall is the first barrier in the defence against heavy metals to avert them from entering the protoplast. Extracellular polymeric substances (EPS) made by the symbionts, including microalgae, are involved in Pb tolerance and ionic biosorption (Pereira et al. 2009; Ozturk et al. 2014). Lacking specialized tissues, the presence of acidic charged polymeric elements around lichen cells may be an important feature for the abstraction and bioaccumulation from the environment of essential oligo elements, most of them being divalent cations (Ca, Fe, Zn, Cu, Mn, etc.) (Hauck et al. 2009b;

Hauck et al. 2010). Some mycobionts present orange or yellow extracellular pigments which belong to the dibenzofuran (e.g. usnic acid, norstictic acid), anthraquinone (e.g. parietin) or pulvinic acid groups which are all efficient UV radiation screens but differ in metal-binding characteristics. Hauck et al. (2009a, 2010) prompt that they may promote adaptation to sites with different pH and metal availability, being complexation a prerequisite for metal uptake control in lichens.

Cyanobacteria and green microalgae can also secret EPS (i.e. proteoglycans and polysaccharides) as a mechanism of protection (Pereira et al. 2009; Casano et al. 2015). The composition of exopolysaccharides varies among species a great deal. They may include uronic acids and pectin-like carbohydrates containing polypeptide moieties forming complex species-specific proteoglycans. Casano et al. (2015) demonstrated a modulation of both moieties in Pb exposed *Trebouxia* axenic phycobionts of *Ramalina farinacea*. Furthermore, the extracellular Pb retention abilities of the microalgae were linked to differences in the chief characteristics of their cell walls and EPS. Therefore, secondary to their role in oligoelement abstraction and bioaccumulation, all these EPS would also be effective in creating microsites that secure cells against toxic heavy metals (Pereira et al. 2009; Casano et al. 2015). In addition to the ability of symbiont cell walls to bioaccumulate Pb, hyperaccumulators, as *Diploschistes muscorum,* have been shown to immobilize it thanks to the excretion of large quantities of oxalates, which form insoluble salts preventing its entrance inside the cytoplasm (Edwards et al. 1997; Sarret et al. 1998).

Regarding thalli anatomy, Goyal and Seaward (1981) reported that the metal accumulation abilities by the different layers of terricolous lichens were interrelated and dependent upon the biologically available metal levels in their associated soils as well as the kind of substrate. Furthermore, metal cumulation ability of the rhizinae was maximum for Pb, Mn and Fe whereas the phycobiont was found to have maximal accumulation capacity for Ni, Cu and Zn. The rhizinae were found to be capable of uptaking, accumulating, translocating and regulating metals. Cations were able to move freely from this area to the upper surface of the thalli and back. The rhizinae and medulla play a significant function in metal management (Goyal and Seaward 1982). Goyal and Seaward (1981) observed that, in the case of *Peltigera* spp., metal accumulation of the algal fraction was consistently higher than that for the fungal fraction except for enhanced Pb levels.

2.3   *Intracellular Effects of Pb*

Despite lichen cell walls are especially well adapted to exclude Pb, it seems that small amounts can still reach the cytoplasm of the symbionts (see Fig. 1). Cd and Cr have been localized in concentric bodies and vacuoles of mycobiont and phycobiont chloroplasts in *Xanthoria parietina* together with other ultrastructural alterations (Sanità Di Toppi et al. 2004; Sanità Di Toppi et al. 2005). However, microscopical studies in *Ramalina duriaei* thalli after experimental exposure revealed that Pb penetrated into the cortical cells of the thallus but not into the algal

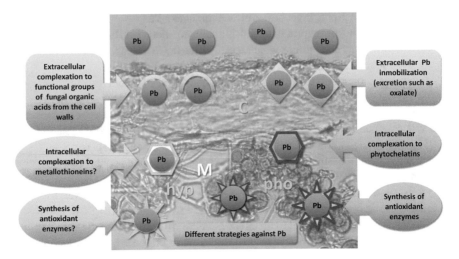

Fig. 1 Mechanisms of tolerance to Pb in lichens. The mechanisms of tolerance to Pb of the myco-biont are marked in yellow and those of the phycobionts in green. When there is more than one phycobiont, these may have different strategies against Pb. Different microalgae are represented with different thicknesses and shade of greens. *C* cortex, *M* mycobiont, *hyp* hyphae, *pho* phycobiont

cells of the phycobiont, nor into the ascospores or medullary cells (Garty and Theiss 1990). Álvarez et al. (2012) reported that *Trebouxia* phycobionts may present extremely diverse abilities to prevent Pb from entering the cytoplasm interfering with cytoplasmatic processes. Whereas *Trebouxia sp.* TR9 tended to accumulate a large amount the heavy metal at the cell wall thus avoiding intracellular uptake, large deposits in small vesicles were seen in *Trebouxia jamesii* (TR1). Purvis and Pawlik-Skowrońska (2008) already highlighted that the microscopical evidence for intracellular localization of metals is very limited in lichens. More works present-ing microscopic direct evidence of Pb cytoplasmic penetration in whole thalli and field samples are necessary to reveal the presumed ultrastructural alterations as well as to check whether the divergences in the literature are due to species-specific strategies.

The synthesis of intracellular chelators such as S-rich peptides and proteins (glu-tathione, phytochelatins and metallothioneins) are well known as the first line of cytoplasmatic defence against heavy metals (Cobbett and Goldsbrough 2002). Pawlik-Skowrońska et al. (2002) showed that widespread lichen species (*Xanthoria parietina*, *Physconia grisea* and *Physcia adscendens*) collected in unpolluted areas biosynthesized phytochelatins and some des-Gly derivatives upon experimental exposure to Cd, Pb and Zn. Mycobionts produced glutathione when grown out of the symbiosis and, conversely, phycobionts (of the genus *Trebouxia*) produced phy-tochelatins (Pawlik-Skowrońska et al. 2002; Bačkor et al. 2006). However, total phytochelatins concentrations were much lower than in other free-living microalgae to similar Cd and Pb concentrations, suggesting that the availability of metals to photobionts is limited as compared to free-living microalgal cells in the presence of the same concentrations of metals (Pawlik-Skowronska 2000; Pawlik-Skowrońska et al. 2002). Perhaps these differences are a consequence of the morphological protection provided by the high mycobiont biomass in lichen thalli in relation to that of the photosynthetic partners.

Metallothioneins have been reported in non-lichenized fungi and most authors postulate an important role for these proteins in lichen tolerance to heavy metals. Nonetheless, no metallothioneins have been identified in lichens in response to Pb or any other metal to our knowledge and this point deserves further investigation.

2.4 Antioxidant Capacity and Redox Dimension of Lichen Symbiosis

Current evidence reveals a neat relationship between metal and oxidative stress in lichens. Both field and laboratory studies have demonstrated a correlation between Pb exposure and lipid peroxidation (Paoli et al. 2011; Gurbanov and Unal 2019). However, the relation between heavy metal exposure/accumulation and peroxidative damage is not direct. As stated above, metals may be scavenged, free radicals may originate from diverse disrupted cellular processes, or else antioxidant defences may counteract prooxidative species.

Thanks to their adaptation to anhydrobiosis, lichens possess especially powerful systems to compensate for oxidative stress generated during rehydration. The capacity to eliminate free radicals, using antioxidants like glutathione, ascorbate, tocopherols and free radical-processing enzymes is important to tolerate desiccation and have a lengthy longevity in the desiccated state (Kranner and Birtic 2005). For desiccation tolerant organisms, rehydration is a critical step. In mitochondria, fast changes in hydric content can alter normal operation and boost reactive oxygen species (ROS) generation. Superoxide anion radicals are of especial concern since it can, in time, be converted to the exceedingly harmful hydroxyl radical by the Haber-Weiss or Fenton reactions catalysed by divalent cations (reviewed in Halliwell and Gutteridge 2007). Free radical production is also increased in these organisms during desiccation and/or rehydration because CO_2 fixation is interrupted, whereas light continues to be absorbed by chlorophyll and electrons transported to O_2 by redox intermediates (Kranner et al. 2008).

Kováčik et al. (2018a, b) observed an increased in reactive oxygen species (ROS) release upon exposure of Cladonia spp. to several heavy metals (Ni, Cu, Cr). However, when Álvarez et al. (2015) challenged Ramalina farinacea with Pb during the critical process of rehydration they observed a strong reduction of the physiological free radical burst together with a change in kinetics in the first hours. The same observation was performed in the isolated axenic Trebouxioid phycobionts of this lichen. Furthermore, lipid peroxidation also was reduced in whole thalli by the effect of Pb. The effects on membrane damage in the isolated phycobionts were diverse. Whereas lipid peroxidation was reduced by high doses of Pb in Trebouxia sp. TR9, no significant effects were observed in *Trebouxia jamesii* (TR1). The authors interpreted these surprising results as a hormetic response to Pb: a compensatory defensive mechanism which would involve the activation of defence systems based on free radical scavenging or antioxidants (Calabrese 2008; Calabrese and Blain 2009; Calabrese 2014). In this sense, together with ROS and lipid peroxidation, they also observed a reduction in chlorophyll auto fluorescence which could be related with the fluorescent quenching ability of lichen microalgae (Wieners et al. 2018). Quenching may help safely dissipate light absorbed by the photosynthetic apparatus preventing free radical formation and photooxidation (Gasulla et al. 2009).

Nevertheless, lichens frequently contain phenolic products with antioxidant and free radical scavenging properties (Kranner and Birtic 2005). A high number of lichen substances chemically related with phenol proved to possess free radical scavenging activity (Valencia-Islas et al. 2007). These authors showed the antioxidant activity and antiradical power of salazinic acid, atranorin, and chloroatranorin (from *Parmotrema stuppeum*) and boninic, 2-O-methylsekikaic, and usnic acids (isolated from *Ramalina asahinae*). At least usnic acid has been shown to increase upon Pb exposure (Gurbanov and Unal 2019) and lichens acclimated to Pb polluted sites showed a high correlation between metal bioaccumulation and lichen substances (Pawlik-Skowrońska and Bačkor 2011).

Ascorbic acid has been shown to be involved in the response and damage reduction of Pb challenged *Coccomyxa subellipsoidea* (a Chlorophyta genus including some lichenizing species) through modulation of ROS/NO balance and metal uptake (Kováčik et al. 2017). *Ramalina farinacea* isolated phycobionts, *Trebouxia sp.* TR9 and *Trebouxia jamesii* (TR1) have been shown to exhibit ascorbate peroxidase enzymatic activity (APx), being higher in TR1 than in TR9. Furthermore, each species differently modulates this activity in response to Pb: whereas it decreased dramatically in the latter, it progressively increased with rising Pb concentrations (Álvarez et al. 2012).

Conversely to plants and free living microalgae, the importance of ascorbate or tocopherol in lichens seems very diminished (Kranner et al. 2005; Purvis and Pawlik-Skowrońska 2008). Whereas it has been demonstrated to decrease in pollution exposed populations of *Parmelia quercina* (Calatayud et al. 1999)*,* no ascorbate or (homologues of) erythroascorbate were found in *Cladonia vulcani* and tocopherol was only present in the microalgae (*Trebouxia excentrica*). Therefore in the absence of these important systems, glutathione plays a pivotal role in oxidative stress prevention, making it essential for lichens to survive desiccation (Kranner and Birtic 2005). In addition to the ability of its cysteine to chelate metals, the tripeptide glutathione is also able to neutralize free radicals directly and is used as co-substrate by a high number of antioxidant enzymes rendering glutathione disulphide (GSSG).

Kranner et al. (2005) proposed that the lichenized symbionts mutually up-regulate their antioxidant systems and when isolated they lose this enhanced resistance to oxidative stress. Outside the holobiont thallus, the alga is very susceptible to photooxidation; without the phycobiont, the fungal glutathione-based antioxidant system seems ineffective. The multifaceted NO has been involved in this redox dimension of lichen symbiosis since scavenging of this molecule was shown to render phycobionts much more susceptible to photooxidation in axenic culture than in the thallus and to be involved in oxidative stress caused by the prooxidant pollutant cumene hydroperoxide (Catalá et al. 2010a, b, 2013). In plants, exogenous sodium nitroprusside supplementation (known to release NO) reduces oxidative damage produced by the presence of Pb (Kaur et al. 2015; Jafarnezhad-Moziraji et al. 2017) and ameliorative effects and regulation of ascorbic acid upon Pb and other heavy metals exposure in *C. subellipsoidea* and *R. farinacea* is linked to NO modulation (Kováčik et al. 2015, 2017, 2019).

Not only enhances the cross talk established between a mycobiont and the phycobionts lichen tolerance to abiotic stress, but is the complexity of symbiosis being revealed, and more than two partners have a starring role in this story. Recent works report that unicellular fungal yeasts, bacteria and a different species of microalgae may render thalli a much more complex system than expected (del Campo et al. 2010, 2013; Casano et al. 2011; Álvarez et al. 2015; Spribille et al. 2016; Moya et al. 2017). *Ramalina farinacea* is sensitive to air pollution (see above), in spite of this, it has been shown to be tolerant to heavy metals such as Pb (Branquinho et al. 1997; Loppi and Frati 2006; Álvarez et al. 2015; Expósito et al. in press). As explained above, two main phycobionts have been identified whose proportions change in dependence of the biogeography of the population studied (Casano et al. 2011). In addition to morphological and genetic differences, both microalgae exhibit very different performances regarding heavy metal and antioxidant defences. Each phycobiont has different ability to immobilize Pb at extracellular level. These two microalgae have completely different strategies to obtain similar levels of Pb tolerance: TR9 has a thicker wall but is less active than TR1 whose enzymes show higher activities constitutively. On the other hand, TR9 is able to upregulate antioxidant defences to achieve similar levels as TR1 upon Pb exposure (del Hoyo et al. 2011; Álvarez et al. 2012, 2015; Casano et al. 2015). Recent results of our group also point to differences in the dependence of each *R. farinacea* phycobiont on NO to maintain the hormetic effect of described above on lipid peroxidation as well as a role in mycobiont-derived NO on Pb response (Expósito et al., in press). This behaviour probably allows them to coexist in the same lichen and amplify the fitness of the organism.

3　Conclusions

Lichens are unique symbiotic phenotypes with complex physiology. They are poikilohydrous, adapted to anhydrobiosis and can resume cell function immediately after rehydrating. Anhydrobiosis and tolerance to desiccation is linked to mechanisms of reparation along with the presence of antioxidants, antioxidant enzymes and NO, among others. In the absence of protective or absorptive specific structures, minerals penetrate thalli and bioaccumulate within cells in a significantly more effective manner than in plants. Heavy metal concentration inside thalli is well correlated with the distance to the pollution source as well as with climatic and land cover factors, greatly impacting on lichen abundance and richness since there is a vast variability in species tolerance. Lichens are good indicators of global change because even subtle environmental changes affect measurably their biodiversity. They are superb passive samplers of atmosphere contamination both for inorganic and organic elements. Namely, they are excellent biomonitors of Pb as a marker for leaded gasoline.

They show a relative high resistance to heavy metals but there exists a debate about the how each symbiont contributes. Pb causes alterations in lichen photosynthesis, membrane damage leading to ion leaking, water loss and decreases of the A-form of DNA involved in replication and transcription processes. However, its effect on the inhibition of enzymes remains to be addressed. Lichen cell walls are especially well adapted to exclude Pb. Extracellular polymeric substances (EPS) made by both myco- and phycobionts are involved in ionic biosorption and Pb tolerance. Hyperaccumulators have been shown to immobilize it thanks to excreted oxalate insoluble salts. Few works present microscopic direct evidence of presumed Pb cytoplasmic penetration in whole thalli and works demonstrating this fact in wild samples are necessary. The ecological success of lichens living on substrates contaminated with toxic elements may be due to the presence of metal-tolerant phycobionts which may present extremely diverse abilities to prevent Pb damage, they biosynthesize phytochelatins in much lower concentrations than in other free-living microalgae. Mycobionts seem to produce glutathione instead, and a presumed synthesis of fungal metallothioneins has not been demonstrated yet. Thanks to their adaptation to anhydrobiosis lichens possess especially powerful systems to compensate for oxidative stress generated during rehydration. *Ramalina farinacea* challenged with Pb during rehydration showed a strong reduction in the physiological free radical burst, a hormetic compensatory mechanism which would involve the activation of defence systems based on free radical scavenging or antioxidants, as well as chlorophyll fluorescent quenching. This ability of lichen microalgae may help safely dissipate light absorbed by the photosynthetic apparatus preventing free radical formation and photooxidation. A high number of lichen substances chemically related with phenol possess free radical scavenging activity and ascorbic acid has been shown to be involved in the modulation of ROS/NO balance and metal uptake. However, the importance of ascorbate or tocopherol in lichens is under debate and some authors point out to a pivotal role of glutathione.

In any case, lichenized symbionts seem to mutually up-regulate their antioxidant systems. Outside the holobiont thallus, the microalga is very susceptible to photooxidation and, without the phycobiont, the fungal antioxidant system seems ineffective. The multifaceted NO has been postulated as a key factor in this redox dimension of lichen symbiosis. The complexity of symbiosis is being revealed and more than two partners have a starring role in this story. Several phycobionts in the same thallus have been identified whose proportions change in dependence of the biogeography of the population studied. Seminal studies show that these coexisting microalgae exhibit completely different strategies to obtain similar levels of Pb tolerance. The lichenization of different partners with diverse physiological strategies and performances seem to be the secret for amplifying the fitness of the organism.

Acknowledgements We wish to thank Dr Jon San Sebastián for the elaboration of Fig. 1. This work was supported by the *Ministerio de Economía y Competitividad* (MINECO-FEDER, Spain) (CGL2016-79158-P) and *Generalitat Valenciana* (GVA, Excellence in Research Spain) (PROMETEOIII/2017/039) and *Comunidad de Madrid - European Commission* (Youth Employment Intiative, Spain) (PEJ-2017-AI/AMB-6337).

References

Agnan Y, Probst A, Séjalon-Delmas N (2017) Evaluation of lichen species resistance to atmospheric metal pollution by coupling diversity and bioaccumulation approaches: a new bioindication scale for French forested areas. Ecol Indic 72:99–110

Agnan Y, Séjalon-Delmas N, Claustres A, Probst A (2015) Investigation of spatial and temporal metal atmospheric deposition in France through lichen and moss bioaccumulation over one century. Science of The Total Environment 529:285–296

Ahmadjian V (1993) The lichen symbiosis. Wiley, New York

Ahmadjian V (1995) Lichens are more important than you think. Bioscience 45:124–124

Álvarez R, del Hoyo A, García-Breijo F, Reig-Armiñana J, del Campo EM, Guéra A, Barreno E, Casano LM (2012) Different strategies to achieve Pb-tolerance by the two *Trebouxia* algae coexisting in the lichen *Ramalina farinacea*. J Plant Physiol 169:1797–1806

Álvarez R, del Hoyo A, Díaz-Rodríguez C, Coello AJ, del Campo EM, Barreno E, Catalá M, Casano LM (2015) Lichen rehydration in heavy metal-polluted environments: Pb modulates the oxidative response of both *Ramalina farinacea* Thalli and its isolated microalgae. Microb Ecol 69:698–709

Arhoun M, Barreno E, Ramis-Ramos G (2000) Releasing rates of inorganic ions in lichens monitored by capillary zone electrophoresis as indicators of atmospheric pollution. Crypt Mycol 21:275–289

Armstrong RA (2017) Adaptation of lichens to extreme conditions. In: Shukla V, Kumar S, Kumar N (eds) Plant adaptation strategies in changing environment. Springer, Singapore, pp 1–27

Aschenbrenner IA, Cernava T, Berg G, Grube M (2016) Understanding microbial multi-species symbioses. Front Microbiol 7:1–9

Augusto S, Máguas C, Matos J, Pereira MJ, Branquinho C (2010) Lichens as an integrating tool for monitoring PAH atmospheric deposition: a comparison with soil, air and pine needles. Environ Pollut 158:483–489

Bačkor M, Loppi S (2009) Interactions of lichens with heavy metals. Biol Plant 53:214–222

Bačkor M, Pawlik-Skowrońska B, Tomko J, Budová J, Sanità di Toppi L (2006) Response to copper stress in aposymbiotically grown lichen mycobiont *Cladonia cristatella*: uptake, viability, ergosterol and production of non-protein thiols. Mycol Res 110:994–999

Bačkor M, Pawlik-Skowrońska B, Budová J, Skowroński T (2007) Response to copper and cadmium stress in wild-type and copper tolerant strains of the lichen alga *Trebouxia erici*: Metal accumulation, toxicity and non-protein thiols. Plant Growth Regul 52:17–27

Bačkor M, Peksa O, Škaloud P, Bačkorová M (2010) Photobiont diversity in lichens from metal-rich substrata based on ITS rDNA sequences. Ecotoxicol Environ Saf 73:603–612

Bačkor M, Péli ER, Vantová I (2011) Copper tolerance in the macrolichens *Cladonia furcata* and *Cladonia arbuscula subsp. mitis* is constitutive rather than inducible. Chemosphere 85:106–113

Bajpai R, Semwal M, Singh CP (2018) Suitability of lichens to monitor climate change. Crypto Biodiver Assess, 182–188

Bargagli R (1998) Trace elements in terrestrial plants: an ecophysiological approach to biomonitoring and biorecovery. Springer, Berlin

Bari A, Rosso A, Minciardi MR, Troiani F, Piervittori R (2001) Analysis of heavy metals in atmospheric particulates in relation to their bioaccumulation in explanted *Pseudevernia furfurea* thalli. Environ Monit Assess 69:205-220

Barreno E (2013) Life is symbiosis. In: Chica C (ed) Once upon a time Lynn Margulis: a portrait of Lynn Margulis by colleagues and friends. Ed. Septimus, Barcelona, pp 56–60

Beck A (1999) Photobiont inventory of a lichen community growing on heavy-metal-rich rock. Lichenologist 31:501–510

Beckett RP, Mayaba N, Minibayeva FV, Alyabyev AJ (2005) Hardening by partial dehydration and ABA increase desiccation tolerance in the cyanobacterial lichen *Peltigera polydactylon*. Ann Bot 96:109–115

Beckett RP, Kranner I, Minibayeva FV (2008) Lichen biology. Stress physiology and the symbiosis. Cambridge University Press, New York

Berganimini A, Scheidegger C, Stofer S, Carvalho P, Davey S, Dietrich M, Dubs F, Farks E, Groner U, Karkkainen K, Keller C, Lokos L, Lommi S, Maguas C, Mitchell R, Pinho P, Richo J, Aragon G, Truscott AM, Wolseley P, Watt A (2005) Performance of macrolichens and lichen genera as indicators of lichen species richness and composition. Conserv Biol 19:1051–1062

Bérmudez GMA, Rodríguez JH, Pignata ML (2009) Comparison of the air pollution biomonitoring ability of three *Tillandsia* species and the lichen *Ramalina celastri* in Argentina. Environmental Research 109 (1):6–14

Bollhöfer A, Rosman KJ (2002) The temporal stability in lead isotopic signatures at selected sites in the Southern and Northern Hemispheres. Geochim Cosmochim Acta 66:1375–1386

Bosch-Roig P, Barca D, Crisci GM, Lalli C (2013) Lichens as bioindicators of atmospheric heavy metal deposition in Valencia, Spain. J Atmos Chem 70:373–388

Branquinho C, Brown DH (1994) A method for studying the cellular location of lead in lichens. Lichenol 26:83–90

Branquinho C, Brown DH, Máguas C, Catarino F (1997) Lead (Pb) uptake and its effects on membrane integrity and chlorophyll fluorescence in different lichen species. Environ Exp Bot 37:95–105

Branquinho C, Catarino F, Brown DH, Pereira MJ, Soares A (1999) Improving the use of lichens as biomonitors of atmospheric metal pollution. Sci Total Environ 232:67–77

Brown DH, Beckett RP (1983) Differential sensitivity of lichens to heavy metals. Ann Bot 52:51–57

Calabrese EJ (2008) Hormesis: why it is important to toxicology and toxicologists. Environ Toxicol Chem 27:1451–1474

Calabrese EJ (2014) Hormesis: a fundamental concept in biology. Microb Cell 1:1–5

Calabrese EJ, Blain RB (2009) Hormesis and plant biology. Environ Pollut 157:42–48

Calatayud A, Abadía Á, Abadía J, Barreno E (1999) Effects of ascorbate feeding on chlorophyll fluorescence and xanthophyll cycle components in the lichen *Parmelia quercina* (Willd.) Vainio exposed to atmospheric pollutants. Physiol Plant 105:679–684

Calatayud A, Deltoro VI, Barreno E, Del Valle-Tascon S (1997) Changes in in vivo chlorophyll fluorescence quenching in lichen thalli as a function of water content and suggestion of zeaxanthin-associated photoprotection. Physiol Plant 101:93–102

del Campo EM, Gimeno J, de Nova JPG, Casano LM, Gasulla F, Breijo FJB, Arminana JR, Barreno E (2010) South European populations of *Ramalina farinacea* (L.) Ach. share different *Trebouxia* algae. In: Nash TH III, Geiser L, McCune B (eds) Biology of lichens: ecology, environmental monitoring, systematics and cyber applications. E. Schweizerbart Science Publishers, Stuttgart, Germany, pp 247–256

del Campo EM, Catalá S, Gimeno J, Del Hoyo A, Martinez-Alberola F, Casano LM, Grube M, Barreno E (2013) The genetic structure of the cosmopolitan three-partner lichen *Ramalina farinacea* evidences the concerted diversification of symbionts. FEMS Microbiol Ecol 83:310–323

Carreras HA, Pignata ML, (2001) Comparison among air pollutants, meteorological conditions and some chemical parameters in the transplanted lichen *Usnea amblyoclada*. Environmental Pollution 111:45–52

Carreras HA, Pignata ML (2002) Biomonitoring of heavy metals and air quality in Cordoba City, Argentina, using transplanted lichens. Environ Pollut 117:77–87

Casano LM, Del Campo EM, García-Breijo FJ, Reig-Armiñana J, Gasulla F, Del Hoyo A, Guéra A, Barreno E (2011) Two *Trebouxia* algae with different physiological performances are ever-present in lichen thalli of *Ramalina farinacea*. Coexistence versus competition? Environ Microbiol 13:806–818

Casano LM, Braga MR, Álvarez R, Del Campo EM, Barreno E (2015) Differences in the cell walls and extracellular polymers of the two *Trebouxia* microalgae coexisting in the lichen *Ramalina farinacea* are consistent with their distinct capacity to immobilize extracellular Pb. Plant Sci 236:195–204

Catalá M, Gasulla F, Pradas del Real AE, García-Breijo F, Reig-Arminana J, Barreno E (2010a) Nitric oxide is involved in oxidative stress during rehydration of *Ramalina farinacea* (L.) Ach. in the presence of the oxidative air pollutant cumene hydroperoxide. In: Nash T III, Geiser L, McCune B (eds) Biology of lichens: ecology, environmental monitoring, systematics and cyber applications. E. Schweizerbart Science Publishers, Stuttgart, pp 87–92

Catalá M, Gasulla F, Pradas del Real AE, García-Breijo F, Reig-Armiñana J, Barreno E (2010b) Fungal-associated NO is involved in the regulation of oxidative stress during rehydration in lichen symbiosis. BMC Microbiol 10:297

Catalá M, Gasulla F, Pradas Del Real AE, García-Breijo F, Reig-Armiñana J, Barreno E (2013) The organic air pollutant cumene hydroperoxide interferes with NO antioxidant role in rehydrating lichen. Environ Pollut 179:277–284

Cernava T, Erlacher A, Aschenbrenner IA, Krug L, Lassek C, Riedel K, Grube M, Berg G (2017) Deciphering functional diversification within the lichen microbiota by meta-omics. Microbiome 5:82

Chettri MK, Sawidis T (1997) Impact of heavy metals on water loss from lichen thalli. Ecotoxicol Environ Saf 37:103–111

Chettri MK, Cook CM, Vardaka E, Sawidis T, Lanaras T (1998) The effect of Cu, Zn and Pb on the chlorophyll content of the lichens *Cladonia convoluta* and *Cladonia rangiformis*. Environ Exp Bot 39:1–10

Chiva S, Garrido-Benavent I, Moya P, Molins A, Barreno E (2019) How did terricolous fungi originate in the Mediterranean region? A case study with a gypsicolous lichenized species. Journal of Biogeography 46:515–525

Cislaghi C, Nimis PL (1997) Lichens, air pollution and lung cancer. Nature 387:463–464

Cobbett C, Goldsbrough P (2002) Phytochelatins and metallothioneins: roles in heavy metal detoxification and homeostasis. Annu Rev Plant Biol 53:159–182

Conti ME, Cecchetti G (2001) Biological monitoring: lichens as bioindicators of air pollution assessment: a review. Environ Pollut 114:471–492

Cumming G, Fidler F, Vaux DL (2007) Error bars in experimental biology. J Cell Biol 177:7–11

Delmail D, Grube M, Parrot D, Cook-Moreau J, Boustie J, Labrousse P, Tomasi S (2013) Halotolerance in lichens: symbiotic coalition against salt stress. In: Prasad M (ed) Ecophysiology and responses of plants under salt stress. Springer, New York, pp 115–148

Dzubaj A, Backor M, Tomko J, Peli E, Tuba Z (2008) Tolerance of the lichen *Xanthoria parietina* (L.) Th. Fr. to metal stress. Ecotoxicol Environ Saf 70:319–326

Edwards HGM, Russell NC, Seaward MRD (1997) Calcium oxalate in lichen biodeterioration FT-Raman spectroscopy. Spectrochim Acta Part A Mol Biomol Spectros 53:99–105

Eldridge DJ, Delgado-Baquerizo M (2018) The influence of climatic legacies on the distribution of dryland biocrust communities. Glob Chang Biol 25:327–336

Ellis CJ, Coppins BJ (2009) Quantifying the role of multiple landscape-scale drivers controlling epiphyte composition and richness in a conservation priority habitat (*Juniper* scrub). Biol Conserv 142:1291–1301

Expósito JR, Coello AJ, Barreno E, Casano LM, Catalá M (2019) Endogenous NO is involved in dissimilar responses to rehydration and Pb(NO$_3$)$_2$ in *Ramalina farinacea* thalli and its isolated phycobionts. Microbial ecology (in press).

Feuerer T, Hawksworth DL (2007) Biodiversity of lichens, including a world-wide analysis of checklist data based on Takhtajan's floristic regions. Biodiversity and Conservation 16: 85–98

Firdous S, Khan S, Dar M, Shaheen H, Habib T, Saifullah T (2017) Diversity and distribution of lichens in different ecological zones of Western Himalayas Pakistan. Bang J Bot 46:805–811

García-Breijo F, Reig-Armiñana J, Salvá G, Vazquez VM, Barreneno E (2010) El liquen *Ramalina farinacea* (L.) Ach. en Asturias. Estructura de talos e identificación molecular de los dos ficobiontes de *Trebouxia* que coexisten. Boletín Ciencias Nat RIDEA 51:325–336

Garty J (2010) Biomonitoring atmospheric heavy metals with lichens: theory and application. CRC Crit Rev Plant Sci 20:309–371

Garty J, Fuchs C, (1982) Heavy metals in the lichen *Ramalina duriaei* transplanted in biomonitoring stations. Water, Air, and Soil Pollution 17:175–183.

Garty J, Theiss HB (1990) The localization of lead in the lichen *Ramalina duriaei* (De Not.) Bagl. Bot Acta 103:311–314

Garty J, Ronen R, Galun M (1985) Correlation between chlorophyll degradation and the amount of some elements in the lichen *Ramalina duriaei* (de not.) Jatta. Environ Exp Bot 25:67–74

Garty J, Karary Y, Harel J (1992) Effect of low pH, heavy metals and anions on chlorophyll degradation in the lichen *Ramalina duriaei* (de not.) bagl. Environ Exp Bot 32:229–241

Garty J, Karary Y, Harel J (1993) The impact of air pollution on the integrity of cell membranes and chlorophyll in the lichen *Ramalina duriaei* (de not.) bagl. transplanted to industrial sites in Israel. Arch Environ Contam Toxicol 24:455–460

Garty J, Cohen Y, Kloog N, Karnieli A (1997) Effects of air pollution on cell membrane integrity, spectral reflectance and metal and sulfur concentrations in lichens. Environ Toxicol Chem 16:1396–1402

Garty J, Tomer S, Levin T, Lehr H (2003) Lichens as biomonitors around a coal-fired power station in Israel. Environ Res 91:186–198

Gasulla F, De Nova PG, Esteban-Carrasco A, Zapata JM, Barreno E, Guéra A (2009) Dehydration rate and time of desiccation affect recovery of the lichenic algae *Trebouxia erici*: alternative and classical protective mechanisms. Planta 231:195–208

Geiser LH, Neitlich PN (2007) Air pollution and climate gradients in western Oregon and Washington indicated by epiphytic macrolichens. Environ Pollut 145:203–218

Giordani P (2007) Is the diversity of epiphytic lichens a reliable indicator of air pollution? A case study from Italy. Environ Pollut 146:317–323

Giordani P, Brunialti G, Alleteo D (2002) Effects of atmospheric pollution on lichen biodiversity (LB) in a Mediterranean region (Liguria, northwest Italy). Environ Pollut 118:53–64

Giordani P, Calatayud V, Stofer S, Seidling W, Granke O, Fischer R (2014) Detecting the nitrogen critical loads on European forests by means of epiphytic lichens. A signal-to-noise evaluation. For Ecol Manage 311:29–40

González CM and Pignata ML (1994) The influence of air pollution on soluble proteins, chlorophyll degradation, MDA, sulphur and heavy metals in a transplanted lichen. Chem. and Ecol. 9:105–113

Goyal R, Seaward MRD (1981) Metal uptake in terricolous lichens:I. Metal localization within the thallus. New Phytol 89:631–645

Goyal R, Seaward MRD (1982) Metal uptake in terricolous lichens: II. Effects on the morphology of *Peltigera canina* and *Peltigera rufescens*. New Phytol 90:73–84

Green TGA, Sancho LG, Pintado A (2011) Ecophysiology of desiccation/rehydration cycles in mosses and lichens. In: Lüttge U, Beck E, Bartels D (eds) Plant desiccation tolerance. Ecological studies (analysis and synthesis), vol 215. Springer, Heidelberg

Gurbanov R, Unal D (2019) The biomolecular alterations in *Cladonia convoluta* in response to lead exposure. Spectrosc Lett 51:563–570

Halliwell B, Gutteridge JMC (2007) Free radicals in biology and medicine. doi: https://doi.org/10.1093/acprof:oso/9780198717478.001.0001

Hasegawa PM, Bressan RA (2000) Plant cellular and molecular responses to high salinity. Annu Rev Plant Physiol Plant Mol Biol 51:463–499

Hauck M, Jurgens SR, Willenbruch K, Huneck S, Leuschner C (2009a) Dissociation and metal-binding characteristics of yellow lichen substances suggest a relationship with site preferences of lichens. Ann Bot 103:13–22

Hauck M, Willenbruch K, Leuschner C (2009b) Lichen substances prevent lichens from nutrient deficiency. J Chem Ecol 35:71–73

Hauck M, Juergens SR, Leuschner C (2010) Norstictic acid: correlations between its physico-chemical characteristics and ecological preferences of lichens producing this depsidone. Environ Exp Bot 68:309–313

Hawksworth DL, Rose F (1970) Qualitative scale for estimating sulphur dioxide air pollution in England and wales using epiphytic lichens. Nature 227:145–148

Herzig R, Liebendorfer L, Urech M, Ammann K, Cuecheva M, Landolt W (1990) Lichens as biological indicators of air-pollution in Switzerland—passive biomonitoring as a part of an integrated measuring system for monitoring air-pollution. Elem Conc Cadasters Ecosyst 35:43–57

Högnabba F (2006) Molecular phylogeny of the genus *Stereocaulon* (*Stereocaulaceae*, lichenized *Ascomycetes*). Mycol Res 110:1080–1092

Honegger R (1991) Functional aspects of the lichen symbiosis. Annu Rev Plant Physiol Plant Mol Biol 42:553–578

Honegger R (1998) The lichen symbiosis—what is so spectacular about it? Lichenologist 30:193–212

del Hoyo A, Álvarez R, del Campo EM et al (2011) Oxidative stress induces distinct physiological responses in the two *Trebouxia* phycobionts of the lichen *Ramalina farinacea*. Ann Bot 107:109–118

Jafarnezhad-Moziraji Z, Saeidi-Sar S, Dehpour AA, Masoudian N (2017) Protective effects of exogenous nitric oxide against lead toxicity in Lemon balm (*Melissa officinalis* L.). Appl Ecol Environ Res 15:1605–1621

Kaur G, Singh HP, Batish DR, Mahajan P, Kohli RK, Rishi V (2015) Exogenous nitric oxide (NO) interferes with lead (Pb)-induced toxicity by detoxifying reactive oxygen species in hydroponically grown wheat (*Triticum aestivum*) roots. PLoS One 10(9):e0138713

Kinalioğlu K, Horuz A, Kutbay HG, Bilgin A, Yalcin E (2006) Accumulation of some heavy metals in lichens in Giresun city, Turkey. Ekologia 25:306–313

Kirk P, Cannon P, David J, Stalpers J (2008) Dictionary of the fungi, 10th edn. CABI Bioscience, United Kingdom

Koch NM, Matos P, Branquinho C, Pinho P, Lucheta F, de Azevedo Martins SM, Ferrao Vargas VM (2019) Selecting lichen functional traits as ecological indicators of the effects of urban environment. Sci Total Environ 654:705–713

Komárek M, Ettler V, Chrastný V, Mihaljevič M (2008) Lead isotopes in environmental sciences: a review. Environ Int 34:562–577

Kováčik J, Klejdus B, Babula P, Hedbavny J (2015) Nitric oxide donor modulates cadmium-induced physiological and metabolic changes in the green alga *Coccomyxa subellipsoidea*. Algal Res 8:45–52

Kováčik J, Rotková G, Bujdoš M, Babula P, Peterková V, Matúš P (2017) Ascorbic acid protects *Coccomyxa subellipsoidea* against metal toxicity through modulation of ROS/NO balance and metal uptake. J Hazard Mater 339:200–207

Kováčik J, Dresler S, Babula P (2018a) Metabolic responses of terrestrial macrolichens to nickel. Plant Physiol Biochem 127:32–38

Kováčik J, Dresler S, Peterková V, Babula P (2018b) Metal-induced oxidative stress in terrestrial macrolichens. Chemosphere 203:402–409

Kováčik J, Dresler S, Micalizzi G, Babula P, Hladky J, Mondello L (2019) Nitric oxide affects cadmium-induced changes in the lichen *Ramalina farinacea*. Nitric Oxide 83:11–18

Kranner I, Birtic S (2005) A modulating role for antioxidants in desiccation tolerance. Integr Comp Biol 45:734–740

Kranner I, Cram WJ, Zorn M, Wornik S, Yoshimura I, Stabentheiner E, Pfeifhofer HW (2005) Antioxidants and photoprotection in a lichen as compared with its isolated symbiotic partners. Proc Natl Acad Sci U S A 102:3141–3146

Kranner I, Beckett R, Hochman A, Nash TH (2008) Desiccation-tolerance in lichens: a review. Bryologist 111:576–593

Laaksovirta K, Olkkonen H, Alakuijala P (1976) Observations on the lead content of lichen and bark adjacent to a highway in Southern Finland. Environ Pollut 11:247–255

Loppi S, Bargagli R (1996) Lichen biomonitoring of trace elements in a geothermal area (central Italy). Water Air Soil Pollut 88:177–187

Loppi S, Frati L (2006) Lichen diversity and lichen transplants as monitors of air pollution in a rural area of central Italy. Environ Monit Assess 114:361–375

Loppi S, Putortì E, Pirintsos SA, De Dominicis V (2000) Accumulation of heavy metals in epiphytic lichens near a municipal solid waste incinerator (central Italy). Environ Monit Assess 61:361–371

Lücking R, Hodkinson BP, Leavitt SD (2016) The 2016 classification of lichenized fungi in the Ascomycota and Basidiomycota approaching one thousand genera. Bryologist 119:361–416

Margulis L, Barreno E (2003) Looking at lichens. Bioscience 53:776

Markert B (1993) Interelement correlations detectable in plant samples based on data from reference materials and highly accurate research samples. Fresenius J Anal Chem 345:318–322

McCune B, Dey J, Peck J, Heiman K, Will Wolf S (1997) Regional gradients in lichen communities of the Southeast United States. Bryologist 100:145–158

Moya P, Molins A, Martínez-Alberola F, Muggia L, Barreno E (2017) Unexpected associated microalgal diversity in the lichen *Ramalina farinacea* is uncovered by pyrosequencing analyses. PLoS One 12(4):e0175091

Nagajyoti PC, Lee KD, Sreekanth TVM (2010) Heavy metals, occurrence and toxicity for plants: a review. Environ Chem Lett 8:199–216

Nash TH III (1996) Lichen biology, 2nd edn. Cambridge University Press, New York

Nash TH III, Gries C (1995) The response of lichens to atmospheric deposition with an emphasis on the Arctic. Sci Total Environ 160–161:737–747

Nash TH III, Gries C (2002) Lichens as bioindicators of sulfur dioxide. Symbiosis 33:1–21

Nimis P, Lazzarin G, Lazzarin A, Skert N (2000) Biomonitoring of trace elements with lichens in Veneto (NE Italy). The Science of The Total Environment 255:97–111

Nimis PL, Castello M, Perotti M (1990) Lichens as biomonitors of sulphur dioxide pollution in la spezia (northern italy). Lichenol 22:333–344

Ochoa-Hueso R, Munzi S, Alonso R, Alonso R, Arróniz-Crespo M, Avila A, Bermejo V, Bobbink R, Branquinho C, Concostrina-Zubiri L, Cruz C, Cruz de Carvalho R, De Marco A, Dias T, Elustondo D, Elvira S, Estébanez B, Fusaro L, Gerosa G, Izquieta-Rojano S, Lo Cascio M, Marzuoli R, Matos P, Mereu S, Merino J, Morillas L, Nunes A, Paoletti E, Paoli L, Pinho P, Rogers IB, Santos A, Sicard P, Stevens CJ, Theobald MR (2017) Ecological impacts of atmospheric pollution and interactions with climate change in terrestrial ecosystems of the Mediterranean Basin: current research and future directions. Environ Pollut 227:194–206

Ockenden WA, Steinnes E, Parker C, Jones KC (1998) Observations on persistent organic pollutants in plants: Implications for their use as passive air samplers and for POP cycling. Environ Sci Technol 32:2721–2726

Ozturk S, Aslim B, Suludere Z, Tan S (2014) Metal removal of cyanobacterial exopolysaccharides by uronic acid content and monosaccharide composition. Carbohydr Polym 101:265–271

Paoli L, Corsini A, Bigagli V, Vannini J, Bruscoli C, Loppi S (2012) Long-term biological monitoring of environmental quality around a solid waste landfill assessed with lichens. Environmental Pollution 161:70–75

Paoli L, Pisani T, Guttová A, Sardella G, Loppi S (2011) Physiological and chemical response of lichens transplanted in and around an industrial area of south Italy: relationship with the lichen diversity. Ecotoxicol Environ Saf 74:650–657

Pawlik-Skowronska B (2000) Relationships between acid-soluble thiol peptides and accumulated Pb in the green alga *Stichococcus bacillaris*. Aquat Toxicol 50:221–230

Pawlik-Skowrońska B, Bačkor M (2011) Zn/Pb-tolerant lichens with higher content of secondary metabolites produce less phytochelatins than specimens living in unpolluted habitats. Environ Exp Bot 72:64–70

Pawlik-Skowrońska B, Di Toppi LS, Favali MA, Fossati F, Pirszel J, Skowronski T (2002) Lichens respond to heavy metals by phytochelatin synthesis. New Phytol 156:95–102

Pereira P, de Pablo H, Rosa-Santos F, Pacheco M, Vale C (2009) Metal accumulation and oxidative stress in *Ulva sp.* substantiated by response integration into a general stress index. Aquat Toxicol 91:336–345

Pinho P, Augusto S, Maguas C, Pereira MJ, Soares A, Branquinho C (2008) Impact of neighbourhood land-cover in epiphytic lichen diversity: analysis of multiple factors working at different spatial scales. Environ Pollut 151:414–422

Pinho P, Dias T, Cruz C, Tang YS, Sutton MA, Martins-Loução MA, Máguas C, Branquinho C, (2011) Using lichen functional diversity to assess the effects of atmospheric ammonia in Mediterranean woodlands. Journal of Applied Ecology 48:1107–1116

Poličnik H, Franc B, Cvetka RL (2004) Monitoring of short-term heavy metal deposition by accumulation in epiphytic lichens (*Hypogymnia physodes* (L.) Nyl.). J Atmos Chem 49:223–230

Puckett KJ (1976) The effect of heavy metals on some aspects of lichen physiology. Can J Bot 54:2695–2703

Purvis OW, Pawlik-Skowrońska B, Cressey G, Jones GC, Kearsley A, Spratt J (2008) Mineral phases and element composition of the copper hyperaccumulator lichen *Lecanora polytropa*. Mineral Mag 72:539–548

Richardson DH (1993) Pollution monitoring with lichens. Richmond Publishing, Slough

Sanità Di Toppi L, Musetti R, Marabottini R, Corradi MG, Favali MA, Badiani M (2004) Responses of *Xanthoria parietina* thalli to environmentally relevant concentrations of hexavalent chromium. Funct Plant Biol 31:329–338

Sanità Di Toppi L, Musetti R, Vattuone Z, Pawlik Skowronska B, Fassati F, Bertoli L, Badiani M, Favali MA (2005) Cadmium distribution and effects on ultrastructure and chlorophyll status in photobionts and mycobionts of *Xanthoria parietina*. Microsc Res Tech 66:229–238

Sarret G, Manceau A, Cuny D, van Haluwyn C, Deruelle S, Hazemann JL, Soldo Y, Eybert Berard L, Menthonnex JJ (1998) Mechanisms of lichen resistance to metallic pollution. Environ Sci Technol 32:3325–3330

Sett R, Kundu M (2016) Epiphytic lichens: their usefulness as bio-indicators of air pollution. Donnish J 3:17–24

Sigal LL, Nash TH, (1983) Lichen Communities on Conifers in Southern California Mountains: An Ecological Survey Relative to Oxidant Air Pollution. Ecology 64:1343–1354

Sipman HJM, Aptroot A (2001) Where are the missing lichens? Mycol Res 105:1433–1439

Sloof JE (1995) Lichens as quantitative biomonitors for atmospheric trace-element deposition, using transplants. Atmos Environ 29:11–20

Spribille T, Tuovinen V, Resl P, Vanderpool D, Wolinski H, Aime MC, Schneider K, Stabentheiner E, Tomme Heller M, Thor G, Mayrhofer H, Johannesson H, McCutcheon JP (2016) *Basidiomycete* yeasts in the cortex of *Ascomycete* macrolichens. Science 353:488–492

Stofer S, Calatayud V, Giordani P, Neville P (2012) Assessment of Epiphytic Lichen Diversity. In United Nations Economic Commission for Europe (Ed.), Manual on methods and criteria for harmonized sampling, assessment, monitoring and analysis of the effects of air pollution on forests (p. 14). Hamburg: UNECE, ICP Forests.

Takala K, Olkkonen H (1981) Lead content of an epiphytic lichen in the urban area of Kuopio, east central Finland. Finn Zool Bot Publ Board 18:85–89

Tehler A, Irestedt M (2007) Parallel evolution of lichen growth forms in the family *Roccellaceae* (*Arthoniales, Ascomycota*). Cladistics 23:432–454

Valencia-Islas N, Zambrano A, Rojas JL (2007) Ozone reactivity and free radical scavenging behavior of phenolic secondary metabolites in lichens exposed to chronic oxidant air pollution from Mexico City. J Chem Ecol 33:1619–1634

Varela Z, López-Sánchez G, Yáñez M, Pérez C, Fernández JA, Matos P, Branquinho C, Aboal JR (2018) Changes in epiphytic lichen diversity are associated with air particulate matter levels: The case study of urban areas in Chile. Ecological Indicators 91:307–314

de Vera JP (2012) Lichens as survivors in space and on Mars. Fungal Ecol 5:472–479

de Vera JP, Alawi M, Backhaus T, Baqué M, Billi D, Böttger U, Berger T, Bohmeier M, Cockell C, Demets R, de la Torre Noetzel R, Edwards H, Elsaesser A, Fagliarone C, Fiedler A, Foing B,

Foucher F, Fritz J, Hanke F, Herzog T, Horneck G, Hübers HW, Huwe B, Joshi J, Kozyrovska N, Kruchten M, Lasch P, Lee N, Leuko S, Leya T, Lorek A, Martínez-Frías J, Meessen J, Moritz S, Moeller R, Olsson-Francis K, Onofri S, Ott S, Pacelli C, Podolich O, Rabbow E, Reitz G, Rettberg P, Reva O, Rothschild L, Sancho LG, Schulze-Makuch D, Selbmann L, Serrano P, Szewzyk U, Verseux C, Wadsworth J, Wagner D, Westall F, Wolter D, Zucconi L (2019) Limits of life and the habitability of Mars: the ESA space experiment BIOMEX on the ISS. Astrobiology 19:145–157

Wieners PC, Mudimu O, Bilger W (2018) Survey of the occurrence of desiccation-induced quenching of basal fluorescence in 28 species of green microalgae. Planta 248:601–612

Zhang Y, Li S, Lai Y, Wang L, Wang F, Chen Z (2019) Predicting future contents of soil heavy metals and related health risks by combining the models of source apportionment, soil metal accumulation and industrial economic theory. Ecotoxicol Environ Saf 171:211–221

Phytoremediation of Lead: A Review

Bhagawatilal Jagetiya and Sandeep Kumar

Abstract Environmental pollution is the most important problem faced by modern civilization among all other concerns. Metals are normal components of the crust of Earth. Due to erosion of rocks, volcanic activity and many more natural and anthropogenic activities metals and other contaminants are discharged and found in almost all environmental compartments and strata. Among these heavy metals, lead is the most considerable toxic pollutant which is coming from diverse sources into the surrounding environment and consequently goes into the various components of the food chain. Industrialisation, urbanization, technological spreading out, increased use of fossil fuel, chemical fertilizer and pesticide use, mining and smelting and inappropriate waste management practices stay put the foremost reasons of extremely high levels of toxic quantities of lead in the environment. Mined ores or recycled scrap metal and batteries are the sources that fulfil the industrial lead requirement. Lead mining-smelting, industrial processes, batteries, colour-paints, E-wastes, thermal power plants, ceramics, and bangle manufacturing are the important point sources of lead. Huge quantities of lead in the air are from combustion of leaded fuel. The key reason for prolonged persistence of lead in the environment is the non-biodegradable character of this metal. This has led to manifold increased levels of lead in the environment and biological systems. Lead has no known biological requirement and is highly toxic even at low concentrations. Lead is looked upon as a strong occupational toxin and its toxicological manifestations are very well documented. Lead toxicity and poisoning has been recognized as a major community health threat all around in developing countries. Lead moves into the ecosystem and creates toxic effects on the microorganism as well as on all living organisms including plants. Conventional or traditional techniques of heavy metal quenching and putting out of contaminants from the contaminated sites have jeopardy to leave go of looming heavy metals in the environment and these are costlier as well as unsafe additionally. Use of microbes and green plants for clean-up purposes is therefore, a promising solution for onslaught of heavy metal polluted sites

B. Jagetiya (✉) · S. Kumar
Phytotechnology Research Laboratory, Department of Botany, M.L.V. Government College, Bhilwara, Rajasthan, India

© Springer Nature Switzerland AG 2020
D. K. Gupta et al. (eds.), *Lead in Plants and the Environment*, Radionuclides and Heavy Metals in the Environment,
https://doi.org/10.1007/978-3-030-21638-2_10

in view of the fact that they include sustainable ways of repairing and re-establishing the natural status of soil and environment. The future outlook of phytoremediation depends on ongoing research and development. The science of phytoremediation has to go through numerous technical obstacles and developmental stages and better outcomes can be achieved by learning and knowing more and more about the variety of biological processes participating in phytoremediation programmes. For successful future of phytoremediation a number of attempts yet to be require with multidisciplinary approach. This review comprehensively presents the background, concepts, technical details, types, strategies, merits and demerits, and upcoming path for the phytoremediation of lead pollution.

Keywords Heavy metals · Lead pollution · Ecotoxicology · Bioremediation · Phytoremediation

1 Introduction

The most important problem faced by modern society today among all other concerns is environmental pollution. Naturally, metals are normal components in soils and in the crust of Earth. Metals are released and are present in various concentrations in different environmental components (water, soil) through a number of discharge processes such as erosion of rocks and volcanic activity. The widespread heavy metals found in contaminated localities are described to be lead, chromium, cadmium, copper, mercury and nickel (Jagetiya and Aery 1994; Jagetiya and Bhatt 2005, 2007; Jagetiya et al. 2007, 2013; Kapourchal et al. 2009; Gupta et al. 2013b). Among these, lead and few other heavy metals are the most significant poisonous and deadly pollutants that come from diverse origin points into the surrounding milieu, plant systems and subsequently come into the food chain. The important lead ore is galena (PbS). Galena has cubic form, low hardness and high density (Reuer and Weiss 2002). Lead has been kept in the category of heavy metal and it is a malleable and soft metal. The average concentration of lead in the soil is about 13 mg kg^{-1} and its values are found to be ranged between 1 and 200 mg kg^{-1}. The ultimate recipient of numerous wastes is soil which comes through various anthropogenic actions, chiefly from mining, industrial discharge and disposal/outflow of wastes from manufacturing, and many more doings. Anthropogenic sources of lead include mining, smelting, electroplating and atmospheric deposition due to petrol and use of pesticides, fertilizers. Atmospheric deposition due to petrol comprises anti-knocking additive lead (Tiwari et al. 2013). Huge quantity of lead into the air entered through combustion of leaded fuel these lead particles then settle down on surface of the soil and goes into the soil with precipitation and irrigation practices. Large sized particles of lead discharged from exhaust of the vehicles by and large go away in the expanse of about 50–100 m from the highways and settled on the surface of soil. On the other hand much farer distance is travelled from these sites by the particles with 2 and less than 2 μm in size (Kapourchal et al. 2009). Soil that contaminated by firing range represent a long term source of lead (Okkenhaug et al. 2016). Hair colouring contributed additionally to lead in environment (permanent colouring has lead acetate combined with SH-group of hair protein to form black insoluble lead sulphides) (Cohen and Roe 1991). Lead mining-smelting, industrial processes, batteries, colour-paints, E-wastes, thermal power plants, ceramics, and bangle manufacturing, etc. are the important point sources of lead pollution (Fig. 1) (Singh et al. 2015). Heavy metal pollution is a worldwide problem because these metals are everlasting and nearly every one of these have lethal and deadly impact on all living being, when their quantity go beyond the threshold limits (Ghrefat and Yusuf 2006; Yadav et al. 2017). Plants as well as animals absorb these toxic heavy metals from surrounding sediments, water, and soils, through ingestion, contact and inhaling of airborne suspended tiny metal particles (Mudgal et al. 2010). Heavy

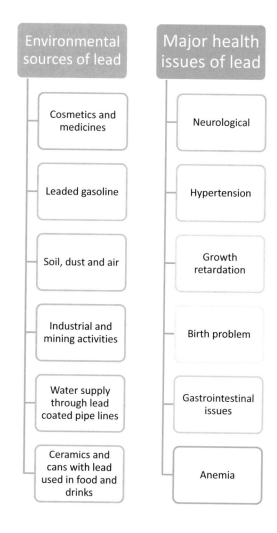

Fig. 1 Major environmental sources and health effects of lead

metals toxicity has reported as the great threat for the health of plant and animals and most of them may disturb important biochemical processes of these organisms. Toxicity due to heavy metals in plants has restraining effect on enzymatic action, stomatal task, photosynthesis, accumulation and uptake of nutrient elements and root system and ultimately on growth (Addo et al. 2012). Heavy metals such as mercury, cadmium and lead do not have any known biological requirement and very much venomous yet at lower concentrations of 0.001–0.1 mg L^{-1}(Aery and Jagetiya 1997; Wang 2002; Alkorta et al. 2004). Cadmium is responsible for carcinogenicity, mutagenicity, endocrine disruptor, lung damage in human (Degraeve 1981; Salem et al. 2000). The major health effect due to mercury toxicity are depression, fatigue, insomnia, drowsiness, hair loss, restlessness, loss of memory, tremors, brain damage, temper outbursts, lung and kidney failure and autoimmune diseases (Neustadt and Pieczenik 2007; Gulati et al. 2010). Excess exposure of lead in kids causes various diseases such as poor intelligence, memory loss, developmental impairment and disabilities in learning, coordination dilemma, and cardiovascular ailment (Fig. 1) (Padmavathiamma and Li 2007; Wuana and Okieimen 2011). Exposure of human beings to these heavy metals that have a number of perilous effects on human health are mostly comes from polluted food chain (Mudgal et al. 2010). A number of heavy metals amputation technologies including ultrafiltration, chemical precipitation, ion-exchange, adsorption, electrodialysis, coagulation-flocculation, reverse osmosis and flotation are generally bring into play. These technologies are too expensive, unfavourable and unsafe to do away with heavy metals from contaminated sites. Above discussed techniques are very costly and has much disadvantage but rather than eco-friendly and cost-effective. Exploiting micro-organisms and plant systems for remediation intentions is therefore a potential way out for pollution due to heavy metal in view of the fact that it includes sustainable decontamination methods to repair and restore the normal state of the rhizosphere and top soil (Jagetiya and Purohit 2006; Jagetiya and Porwal 2019; Jagetiya and Sharma 2009, 2013; Jagetiya et al. 2011, 2012, 2014; Yadav et al. 2017). The efficient and most attractive alternative is plant based remediation or phytoremediation which has already been used of years is environment pleasant, inexpensive and safer modus operandi. It has minimal vicious impact on the ecosystem (Kapourchal et al. 2009; Singh 2012; Ali et al. 2013). A large number of studies have been successfully carried out for phytoremediation of lead and number of plant species are being used for this purpose (Prasad and Freitas 2003; Kapourchal et al. 2009; Malar et al. 2014; Arora et al. 2015; Mahar et al. 2016; Wan et al. 2016; Fanna et al. 2018; Chandrasekhar and Ray 2019).

2 Lead Enrichment in the Environment

Lead is present in Earth's crust as a bluish-grey, low melting, heavy metal and it is exceptionally found as a natural metal and present frequently combined with two or more elements to constitute compounds of lead (ATSDR 2005). It is one of the metals usually present in the environment for the reason that it is in the list of the earliest discovered metals and most far and wide utilized in history of human beings (Shoty et al. 1998). It is becoming severe menace to human health in view of the fact that it's continued to go into the environment as an automobile exhaust emission and widespread exploitation in industry (Juberg et al. 1997). Mined ores (primary) or recycled scrap metal and batteries (secondary) are the sources of lead used in industries. It is reported that about 97% of lead-acid batteries are recycled and lead predominantly found nowadays is "secondary" type and accrued from lead-acid batteries (ATSDR 2005). Manufacturing of lead batteries, extensively used in automobiles is the main use of lead in the industries. Lead is also used for shielding of X-ray machines, alloy making, manufacturing of corrosion and acid resistant stuffs and soldering materials manufacturing etc. (Patil et al. 2006). Lead pollution in air, water, soil and agricultural fields is an ecological concern due to its severe impact on human health and environment since among heavy metals lead is most hazardous. Mining-smelting, industrial effluents, fertilizers, pesticides, and municipal sewage sludge are the main sources of lead pollution in the environment (Aery et al. 1994; Sharma and Dubey 2005; Malar et al. 2014). Negatively charged solid surfaces such as clays, carbonates, oxides and hydroxides of iron, manganese as well as organic carbon of water column rapidly scavenge soluble lead. Consequently non-chelated/dissolved lead has a short water column dwelling duration in ocean. Settling of lead associated particulate stuff by and large regulates the distribution of lead in specific ocean basin (Chakraborty et al. 2015). In maritime sediments, lead may be found in diverse physico-chemical varieties and it has differential affinities for various binding-phases of coastal sediments. Carbonate phase in coastal sediments plays an imperative role in regulating lead distribution (Fulghum et al. 1988) and scavenging nature of lead by Fe/Mn oxy-hydroxide phase in residue has also been identified as a crucial process (Jones and Turki 1997). Distribution and speciation of lead has been demonstrated to be regulated by organic binding phase of it (Krupadam et al. 2007; Chakraborty et al. 2012). Geogenic or anthropogenic activities turned lead contamination into a severe large-scale worldwide environmental apprehension. Industrialisation, uncontrolled use of fossil fuel resources, urbanization, technological expansion, use of pesticides and fertilizers, mining and smelting and poor waste management are the foremost reasons of extremely high quantities of lead in the environment (Lajayer et al. 2017; Chandrasekhar and Ray 2019). Enormous mining activities, paper, metal coating, fertilizer and other industries resulted in the diffusion of lead and allied heavy metals into the environment and their concentrations is escalating bit by bit (Fu and Wang 2011; Wang et al. 2016). Increased lead levels in the water reservoirs is taking place due to residential dwellings, groundwater infiltration, mining drains and manufacturing discharges and over the most recent years, growing human population and industrial expansion have led to a boost of lead contamination in aquatic ecosystems. For that reason, studies reporting the effects of lead and other toxic heavy metals on aquatic organisms are presently attracting added contemplation, predominantly those focusing on urban and industrial contamination (Rocchetta et al. 2007; Akpor and Muchie 2011; Sadik et al. 2015; Dogan et al. 2018). The blemish of coastal waters with trace and heavy metals through anthropogenic spring and sewage has turn into a ruthless predicament (Mamboya et al. 1999). Heavy metals, such as lead is among the most widespread pollutants at hand in equal amounts in urban and industrial discharge (Sheng et al. 2004; Santos et al. 2014). Environmental degradation from heavy and toxic metal contaminants in aqueous water streams and groundwater as well as in soil posing a major community problem is mainly due to worldwide technological progress, unprecedented anthropogenic activities (over exploitation of metal-mineral resources, over use of fertilizers and pesticides, increased household activities and automobiles exhaust) and natural phenomenon (forest fires, volcanic eruption and seepage from rocks) that needs to be addressed seriously. Heavy metals and minerals especially lead, mercury, chromium, cadmium, copper, arsenic and aluminium is a serious threat to the environment and human health. These toxic substances enter into the human body mainly through contaminated water, food and air, leading to numerous lethal health complications (Singh et al. 2015). Some other reports also states that sources of heavy metals in the environment are mainly industry, municipal wastewater, atmospheric pollution, urban runoff, river dumping, and shore erosion and stated that anthropogenic inputs of metals exceeds natural inputs. Higher volumes of cadmium, copper, lead and iron may be act like ecological poisons in terrestrial and aquatic ecosystems (Balsberg-Påhlsson 1989; Guilizzoni 1991). The water, sediments and plants in water bodies receiving municipal and domestic runoff contain higher quantity of heavy metals in comparison to those not getting runoff from urban areas and this process leads into surplus metal levels in surface water which cause a health risk to human beings and to the environment both (Vardanyan and Ingole 2006). Higher levels of lead in the forest flooring and relatively porous soils in forest ecosystems has been documented that lead is released from the forest flooring to the mineral soil or into the surface waters. Continued accrual of lead in forest ecosystems consequently may pose upcoming threat to water quality (Johnson et al. 1995). Lead reaches to the soil and environment through pedogenic processes (depends on the nature and origin of the parent substances) and through anthropogenic activities. Anthropogenic processes, primarily involve manufacturing activities and the disposal of industrial and municipal waste materials and these are the major source of lead contamination of environment (Adriano 2001). Foremost important sources of lead enrichment in the environment are presented in Fig. 2.

3   Ecotoxicology of Lead

Lead is one of the earliest metals discovered by the human and its distinctive nature, such as pliability, ductility, higher malleability, low melting point and corrosion resistant, make its widespread usages in numerous industrial process (colour-paint, automobiles, plastics, and ceramics). The key cause for long-lasting persistence of lead in the ecosystem is due to its non-biodegradable character; consequently it has led to a manifold quantity of free lead in the environment and living beings. Lead is considered as a powerful/potent occupational pollutant and its toxicological

Fig. 2 Various sources of lead enrichment in the environment

manifestations are very well recognized. Lead poisoning has been documented as a major public and community health peril for the most part in developing countries of the world. Nevertheless, a variety of community health and occupational approaches have been taken on in order to control and regulate the lead toxicity, many more cases of lead poisoning are yet to be accounted (Flora et al. 2012). Diverse sources together with industrial activities including smelting of lead and coal burning, colour-paints containing lead, pipes having lead or lead based soldering in water supplying system, recycling of batteries, bearings and grids, lead-based gasoline, etc. are the reasons for human exposure to lead. Though lead toxicity is a decidedly explored and meticulously published topic, full control and preclusion concerning on exposure to lead is yet far from being accomplished. Lead is a non-essential element and has no advantage on to the biological systems and no "safe" level of exposure to lead has been reported. There is even no such concentration of lead is reported to found essential for it to require by living beings and toxicity of lead have been reported as specific menacing hazard with the potential of causing irreversible health consequences. Lead moves into and throughout the ecosystem and creates toxic effects on the microorganism and all living organisms. It is a highly toxic heavy metal that affects human beings, animals, plants and phytoplankton by incorporating into food chain (Truhaut 1977; Chapman 2002; Huang et al. 2011; Singh et al. 2012).

3.1 Effects of Lead on Living Beings and Human Health

No function of lead is known for biological systems, likewise it causes many irreversible health problems once it taken up in the tissues of living systems. Lead toxicity in the environment is an ancient and continual community health concern for all the countries of the world. All the important organs such as hematopoietic, renal, nervous and cardiovascular systems are affected by lead toxicity. Oxidative stress has been reported as pronounced and severe effect of lead toxicity. Biomolecules such as enzymes, proteins, membrane lipids and DNA are damaged by excess lead toxicity which is responsible for generating ROS that impairs the antioxidant defence system (Fig. 3) (Flora et al. 2012; Inouhe et al. 2015). All the way through the evolutionary process lead incorporates into the tissues of living organisms and thus has

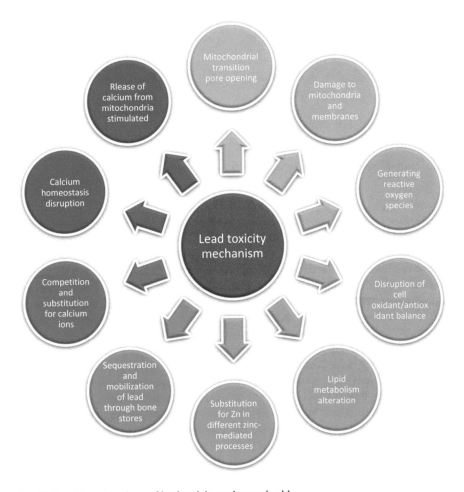

Fig. 3 Possible mechanisms of lead toxicity on human health

become crucial element. Lead particles may enter into residential houses through windows, shoes, air, and so on (Patel et al. 2006). Improvement of technology over the time has appreciably decreased the discharge of lead but still conditions at local sites may be present causing a potential risk due to exposure in surrounding environment. Children are more susceptible to lead toxicity because of their activities of hand to mouth, high rates of respiration and additional absorption by gastrointestinal systems per unit body weight. For certain at-risk groups of children lead toxicity continues to be an main community health issue and impacts of lead on intellectual development has been remained a major concern forever (Ahamed and Siddiqui 2007). Lead affects the nervous system of vertebrates and cause diseases in fingers, wrists or ankles. Accumulation of lead in invertebrates above a particular level becomes toxic to their predators. Elevated blood levels of lead just $>10 \mu g \, dL^{-1}$ cause anaemia in children (Tiwari et al. 2013). Two types of anaemia reported, haemolytic anaemia and frank anaemia due to lead poisoning effect on the enzyme δ-aminolevulinic acid dehydrates (ALAD), aminolevulinic acid synthetase (ALAS), ferrochelatase involved in haem synthesis but mainly affects the cytosolic enzyme ALAD. Lead nitrate induce the rate-limiting enzyme ALA synthetase (S-aminolevulinic acid synthetase) of the haem biosynthesis at the post transcriptional level (Kusell et al. 1978). Lead exposure had decreased the permanency of the spermatozoa and reduced secretory function of the accessory genital glands (Wildt et al. 1977). Nutrients factors or deficiency (some vitamins and essential elements) affects the susceptibility to lead toxicity (Ahamed et al. 2005). Renal effects of Pb poisoning ($>60 \mu g \, dL^{-1}$) causes Fanconi's syndrome which is represented by the combined excretion of phosphates, glucose and amino acids at an abnormal rate. It also affects the ROS production and antioxidant defence, causing cell death due to oxidative stress (Gupta et al. 2009; Flora et al. 2012). A variety of immune responses such as differentiation of B cell, MHC class II molecule on the surface of B lymphocytes' increased expression, lymphocyte proliferation, inhibition and targeted first suppressor T cell and then Th cells are the poisonous effects of lead. Lead also affects the nuclear factor-kb, CD 4, natural killer cells (NKC) and nitric oxide (Singh et al. 2003). Lead may cause cytotoxicity and genotoxicity, which can be determined through histopathology, proteomics and cell growth (Pan et al. 2010).

3.2 Effect of Lead on Microorganisms

Elevated level of lead near smelter decrease the microorganism population and inhibited the germination of fungal spore and mycelium growth (Bisessar 1981). Lead toxicity turn into inhibited cell division, protein denaturation, cell membrane disruption, inhibition of enzyme activity, translation inhibition and transcription inhibition by damaging of DNA (Yadav et al. 2017). Pb caused short term (<24 h, 5 mg L^{-1}) and long term impact on microorganisms. After long term lead exposure on sludge bacterial viability decreased linearly (Yuan et al. 2015). E-waste recycling sites produce different lead contaminants that effects the soil microorganisms by decrease biomass, enzyme activity (covalently bind with –SH, –OH, –COOH, –NH$_2$ group on active site of enzymes) and enhance soil basal respiration, metabolic quotient (Zhang et al. 2016). Lead affects the algae by inhibiting growth and primary metabolite accumulation (Piotrowska et al. 2015). Lead accumulations in zooplankton were lower than in bacteria and in phytoplankton (Rossi and Jamet 2008).

3.3 Effects of Lead on Ecosystem

Industrial fine particles are toxic for ecosystem (Schreck et al. 2011). Lead stored in the O horizon of soil in forest floor from input deposition of alkyl-lead additives gasoline due to this lead release in the mineral soil or surface water from forest floor with high concentration threat to water quality in ecosystem (Johnson et al. 1995). Acute/chronic toxicity concentration or threshold concentration of dissolved lead in fresh water ecosystem is calculated between 6.3 mg dissolved Pb L^{-1} and 31.1 mg dissolved Pb L^{-1}. It is influenced by many reported factors effects of pH, alkalinity, dissolved organic carbon and concentration of Mg and Ca cations in aquatic toxicity of lead at EU scenarios (Sprang et al. 2016). Some widely used applications like UV coating, polishers and paints used Meo and NPs species that release cerium oxide (CeO$_2$NPs) which deposit on aquatic sediments and therefore making potential risk of lead in aquatic ecosystem (Wang et al. 2018). Lead toxicity affects aquatic organisms such as blackening in the tail and spinal deformity (Singh et al. 2012). Some physiological and biochemical changes have also been reported in hydrophytes such as water hyacinth (Malar et al. 2014).

3.4 Ecotoxicology of Pb-210

Pb-210 is the most important to study in relation to its behaviour in soils and plants. Short-lived progeny of radon decay gives rise to ^{210}Pb. This radioactive isotope exhibits the chemical characteristics of lead because it has sufficient time to decay, which results in the production of Po-210 which tends to be mobile in the substratum. The most common redox state encountered in the environment is the divalent form of lead, out of the three known oxidation states, 0, +2 and +4. Lead is found adsorbed on the surface of oxides, hydroxides, oxyhydroxides, clays and organic matter. The adsorption is highly associated with the cation exchange capacity of and pH of the soils. Phosphate, chloride, and carbonate and other soil constituents affect lead reactions in the soils by precipitation and reducing adsorption due to ligand forming (Mitchell et al. 2013). Pb-210 produced in the atmosphere from ^{222}Rn and this phenomenon increases the concentration of ^{210}Pb with decrease in Ra-226. Pb-210 will decay to give rise ^{210}Po (Sheppard et al. 2008). A major contributing way to plant uptake of ^{210}Pb is aggregating from surrounding atmosphere (Ham et al. 2001). Po-210 to Pb-210 equilibrium studies established the trail of ^{210}Pb settling. Po-210 to Pb-210 ratios less than 1 demonstrate inadequate time for ^{210}Po to equilibrate subsequent uptake by plants. The ratio for shoots was found to be between 0.35 and 0.72 during a study (Pietrzak-Fils and Skowronska-Smolak 1995). Canadian annual plants showed a median value of 0.6 (Sheppard et al. 2004, 2008). Relative concentrations of radionuclides in the medium, individual Fv values, rate of deposition of radionuclides on the shoots from atmosphere and further withholding determined the ratio. Higher Pb-210 Fv values were noticed relative to stable lead due to atmospheric deposition (Sheppard et al. 2004). In contrast, an excess of ^{210}Po over ^{210}Pb was observed in wild berries from a boreal ecosystem (Vaaramaa et al. 2009). To look into phytoremediation of lead a vibrant model developed that gave a mechanism of lead behaviour within soil–plant system (Brennan and Shelley 1999). Limited studies are available for radio-ecological research (Hovmand et al. 2009; Sheppard et al. 2008; Vaaramaa et al. 2009). Atmospheric settling of ^{210}Pb was found the major route under a study when uptake under covered tent was compared to that in open field (Pietrzak-Fils and Skowronska-Smolak 1995). The effects of soil texture on transfer were the highest when plants were grown on sandy soils. There were few noteworthy differences between crop groups, and no correlations were found between numerous soil characteristics (cation exchange capacity, pH, clay and organic matter) and the Fv values for the crop sets. Significant dissimilarity was noticed in Fv values among soil types for leafy vegetables, root crops and tubers (Vandenhove et al. 2009). The Fv value may be increased up to 20-fold from atmospheric settlement of ^{210}Pb (straw of cereals, grasses and vegetables) (Vandenhove et al. 2009). The accumulation of lead and other radionuclides in spring wheat exhibited the following relationship: root > stem > grain (Nan and Cheng 2001), while the beet of red beet had a lower value than the leaves (Pietrzak-Fils and Skowronska-Smolak 1995). On the other hand, in beans ^{210}Pb was largely absorbed and held in the roots without translocation to aerial plant parts (D'Souza and Mistry 1970). Red kidney bean showed that 100% was retained by the leaf with an application of ^{210}Pb as nitrate to the leaves. This has been known as immobile isotope and trapped at the sites of applications (Athalye and Mistry 1972). Type of the plant and part of the plants plays important role for plant ^{210}Pb content (Pietrzak-Fils and Skowronska-Smolak 1995). In crops grown under ordinary field conditions, washing may take away about 10% of plant radioactivity; radioactivity values were found 6–10-fold high in plants grown in the open in contrast to crops kept in the cover-up tents. During dry time radioactivity of ^{210}Pb on plant leaves was found at climax, while during wet times it was observed to be decreased and attributed that during wet of aerosols wash-off from the surface of the leaves (Sugihara et al. 2008; Mitchell et al. 2013). Transfer factors of ^{210}Pb from contaminated soil in oil fields located in a semiarid area to some pasture species were determined and it was found that uptake of ^{210}Pb from soil to plants increased with the time of the first planting. Among the studied plants *Medicago sativa* (alfalfa) and Bermuda grass were found to have the highest transfer factor (Al-Masri et al. 2014). In *Typha latifolia* L. in a study conducted in an environment with a higher quantities of radionuclides and heavy metals, many structural alterations; synthesis and presence of numerous antistress substances (anthocyanin, ferritin, etc.) as well as the occurrence of various exogenous particles in the epidermal and parenchyma cells were observed (Corneanu et al. 2014).

4   Remediation Techniques of Lead

The method in which contaminants from soils, water and air are removed is known as remediation. Heavy metal such as lead is one of the most dangerous contaminant. Electrokinetic remediation (EKR) uses many electrolytes to bind contaminants and make them immobile in soil by the influence on soil conductivity (H^+ and Fe^{2+}) and current by replaced soil ions from EKR ions (e.g., KNO_3, $NaNO_3$, Na_2CO_3, K_2HPO_4, KH_2PO_4, sodium acetate acid (NaAc), H^+, EDTA, Na/HAc, citric acid, Tris–acetate-starch, ammonium, nitrate, lithium lactate, $MgSO_4$, and NH_4NO_3) (Li et al. 2014).

4.1   Conventional Remediation Techniques

In situ vitrification, excavation and landfill, soil incineration–washing–flushing–reburial, solidification, stabilization of electrokinetic system, pump and treat system, ion exchange chemical precipitation, ultrafiltration, adsorption, electrodialysis, flocculation, and so on are mostly used decontamination methods for metal polluted sites; out of these, ion exchange, adsorption, ultrafiltration, chemical precipitation, electrodialysis, and flocculation are more useful for lead removal.

Chemical precipitation: Coagulants such as lime, alum and iron salt are used for precipitation of metal ions.
Ion exchange: Electrostatic force on ion exchange in a dilute solution is applied.
Adsorption: A molecular or atomic film is formed by accumulation of gas or liquid solutes on the surface of an adsorbent.
Ultrafiltration: It is used to remove heavy metal ions; 0.1–0.001 micron pore size membrane used in ultrafiltration.
Eletrodialysis: This method is applied when separation of cations and anions through electrical potential to remove metal ions by the use of semipermeable ion selective membranes.
Flocculation: This method makes flocs in water using a coagulant to attract suspended metal ions by these flocs (Yadav et al. 2017).

These methods have a threat of releasing potentially dangerous metals into the environment as well as unsafe, high-priced and inadequate.

4.2   Bioremediation Techniques

Use of microbes and green plants for clean-up purposes is therefore, a promising solution for onslaught of heavy metal polluted sites in view of the fact that they comprise sustainable ways of repairing and re-establishing the natural status of soil and environment. The employment of primarily microbes, to clean up contaminated soils, aquifers, sludge, residues and air, termed as "bioremediation", is a rapidly

changing and expanding branch of environmental biotechnology that offers a potentially more effective and economical clean up method. The use of microorganisms
to control and destroy toxic substances is of growing attention to minimize a number
of pollution issues. Bacteria, algae, fungi and yeast and some other microbes have
been found to absorb and break down many metal compounds (Dixit 2015). Green
plants may be used to remove effluents and contamination from soil. This may be
called as "phytoremediation" (Jagetiya et al. 2011, 2014; Gupta et al. 2013a, b).

4.2.1 Remediation of Lead by Bacteria

Many bacterial species accumulate lead from polluted soil and water system by the
process of bioaccumulation and bio-sorption through active and passive process. To
survive in the toxic environment these species develop resistance to toxicity of
heavy metals. Some potential bacterial species being used for lead remediation are
listed in Table 1.

4.2.2 Remediation of Lead by Algae

Algae remove lead by the process of chemisorption in which metal ion transport
into cytoplasm and physical adsorption in which ion adsorbed over the surface
quickly (Dwivedi 2012). The mechanism of remediation depends on anatomy of
algae and environmental conditions in growing medium (Yadav et al. 2017). In
recent years many researchers have used various algal species for removal of lead
from contaminated sites (Table 2).

4.2.3 Remediation of Lead by Fungi

Comparatively fungi are the good alternative for removal of heavy metal from the
environment and more tolerant to heavy metals than the bacterial species (Rajapaksha
2004). Some fungi work as hyper-accumulator of heavy metals (Purvis and Halls

Table 1 Some bacterial species with potential of Pb bioremediation.

Bacterial species	References
Bacillus firmus	Salehizadeh and Shojaosadati (2003)
Bacillus licheniformis	Basha and Rajaganesh (2014)
Corynebacterium glutamicum	Choi and Yun (2004)
Escherichia Coli	Basha and Rajaganesh (2014)
Pseudomonas aeruginosa	Lin and Lai (2006)
Pseudomonas fluorescens	Basha and Rajaganesh (2014)
Pseudomonas putida	Uslu and Tanyol (2006)
Salmonella typhi	Basha and Rajaganesh (2014)

Table 2 Some algal species with potential of Pb bioremediation

Algae species	References
Ascophyllum nodosum	Holan and Volesky (1994)
Chlorella vulgaris	Aung et al. (2012); Edris et al. (2012)
Cladophora fascicularis	Deng et al. (2007)
Cladophora glomerata	Dwivedi et al. (2012)
Oedogonium rivulare	Dwivedi et al. (2012)
Oscillatoria quadripunctulata	Rana et al. (2013); Azizi et al. (2012)
Oscillatoria tenuis	Ajavan et al. (2011)
Sargassum natans	Holan and Volesky (1994)
Sargassum vulgare	Holan and Volesky (1994)
Spirogyra hyalina	Kumar and Oommen (2012)

Table 3 Some fungal species with potential of lead bioremediation

Fungal species	References
Aspergillus niger	Kapoor et al. (1999)
Aspergillus flavus	Dwivedi et al. (2012)
Aspergillus terreus	Joshi et al. (2011); Massaccesi et al. (2002)
Mucor rouxii	Yan and Viraraghavan (2001)
Saccharomyces cerevisiae	Damodaran et al. (2011)
Saprolegnia delica	Ali and Hashem (2007)
Trichoderma viride	Ali and Hashem (2007)

1996). Cell wall lipids, carbohydrates and proteins bind with the metals (Veglio and Beolchini 1997; Beolchini 2006). Potential fungal species used for lead remediation are given in Table 3.

4.2.4 Phytoremediation (Green Technology)

When conventional remediation methods are unfeasible due to the extent of the polluted region or cost and safety issues, phytoremediation is advantageous (Garbisu and Alkorta 2003). Phytoremediation involves different methods where green plants efficiently decontaminate polluted sites at relatively low cost and good public acceptance. Phytoremediation is an aesthetically pleasing, safer and non-destructive, sustainable technology which has commercial acceptability (Sheoran et al. 2011). In this modern technology accumulation power of plants is used to detoxify essential and non-essential heavy metals from contaminated soils (Djingova and Kuleff 2000). Some most important families of plant that have been identified to accumulate heavy metals are Fabaceae, Euphorbiaceae, Asteraceae, Brassicaceae, Lamiaceae and Scrophulariaceae and mangrove plants (Lacerda 1998). Plants used in phytoremediation accumulate toxic heavy metal in varied concentrations at same

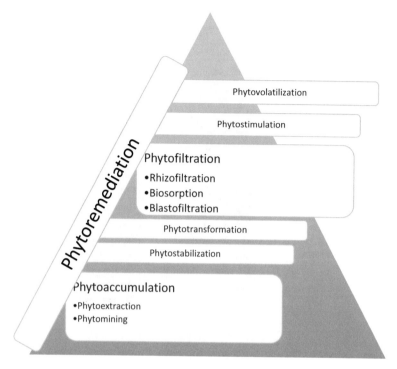

Fig. 4 Various methods of phytoremediation

contaminated site and some plant work as hyper accumulators that absorb 100-fold greater amount than those of non-accumulator plants of heavy metals (Peer et al. 2005). Toxic heavy metals from underground water, dregs, top soil, and brown fields can be removed by various methods of phytoremediation (Fig. 4). Reported values of conventional remediation technologies are always higher than the phytoremedial techniques which is commercially applicable and having all adequate possibilities to be applied successfully. Mostly around the world phytoremediation studies are confined only to the organic chemistry and bio-agro processes. However, individual monetary and financial analysis for this process is largely unavailable (Ali et al. 2013). Economics of phytoremediation consists of two types of costs, that is, initial capital and running or operational costs. The mandatory materials involved in initial capital may be pollution analysis and preliminary testing, planning and setting up of decontamination or removal tactics, preparation of soil, nursery tools (quantitative) procurement, creation of storing facility, irrigation facility, incineration utensils and apparatus, road construction, bridge construction, and drain facility. Operational cost mostly have the cost of ploughing, seedling, plantation programmes, irrigation, fertilizers-pesticides-insecticides-herbicides purchase and application, produce harvesting with a number of less considerable things. Benefit of cost includes both of benefits during remediation and after remediation. Therefore, phytoremediation

Table 4 Some higher plant
species with potential of Pb
phytoremediation

Plant species	References
Alyssum lesbiacum	Baker et al. (1991)
Alyssum murale	Baker et al. (1991)
Ambrosia artemisiifolia	Huang and Cunningham (1996)
Arabidopsis thaliana	Baker et al. (1991)
Astragalus bisulcatus	Baker et al. (1991)
Brassica juncea	Kumar et al. (2002)
Brassica oleracea	Baker et al. (1991)
Euphorbia cheiradenia	Chehregani and Malayeri (2007)
Jatropha curcas	Abhilash et al. (2009); Jamil et al. (2009)
Populus deltoides	Ruttens et al. (2011)
Populus nigra	Ruttens et al. (2011)
Populus trichocarpa	Ruttens et al. (2011)
Raphanus sativus	Baker et al. (1991)
Thlaspi caerulescens	Baker et al. (1991)
Trifolium alexandrinum	Ali et al. (2012)
Zea mays	Huang and Cunningham (1996); Meers et al. (2010)

technology is price efficient compared to conventional remediation technologies (Wan et al. 2016). Some potential plant species used for lead phytoremediation are listed in Table 4.

Phytoaccumulation

Translocation and uptake of metal from contaminated soil or water by plant root and accumulation in above ground biomass is the basic concept of phytoaccumulation and this has been described in literature as many other terms such as phytoabsorption, phytoextraction and phytosequestration (Chou et al. 2005; Eapen et al. 2006; Singh et al. 2009). Metal accumulation in shoot is an effective biochemical process (Zacchini et al. 2011). Natural/continuous or induced (driven by chelators) are the two techniques of phytoextraction (Hseu et al. 2013). Certain plants work as hyperaccumulators because of having 100 times more absorbing power (Table. 5). In *Zea mays,* which is a high biomass crop, can accumulate higher lead in shoots than in roots (Brennan and Shelley 1999; Gupta et al. 2009) whereas, plants such as *Thalspi rotundifolium* are low biomass plants can hoard lead higher in roots than shoots. For better performance plant should have stumpy growth rate, elevated production of biomass, property of hyper-accumulation of metals or contamination to be removed, branched roots, higher root to shoot translocation of heavy metals, high tolerance and adaptability, pests and pathogen resistance, easy to cultivate and harvest (Adesodun et al. 2010; Sakakibara et al. 2011). Phytomining is a type of phytoextraction that involves extracting of heavy

Table 5 Hyperaccumulator plant species for phytoextraction of lead

Plant	Family	References
Arabis paniculata	Brassicaceae	Tang et al. (2009)
Noea mucronata	Amaranthaceae	Chehregani et al. (2009)
Baccharis latifolia	Asteraceae	Bech et al. (2012)
Onchus oleraceus	Asteraceae	Bech et al. (2012)
Bidens triplinervia	Asteraceae	Bech et al. (2012)
Brassica juncea	Brassicaceae	Zaier et al. (2010)
Buckwheat	Polygonaceae	Chen et al. (2004)
Cynara cardunculus	Asteraceae	Epelde et al. (2009)
Helianthus annuus	Helianthoideae	Chen et al. (2004)
Hemidesmus indicus	Apocynaceae	Sekhar et al. (2005)
Lepidium bipinnatifidum	Brassicaceae	Bech et al. (2012)
Indian mustard	Brassicaceae	Chen et al. (2004)
Poa pratensis	Poaceae	He et al. (2009)
Pisum sativum	Fabaceae	Chen et al. (2004)
Plantago orbignyana	Plantaginaceae	Bech et al. (2012)
Sedum alfredii	Crassulaceae	Gupta et al. (2010)
Sesuvium portulacastrum	Aizoaceae	Bech et al. (2012)
Sonchus oleraceus	Asteraceae	Xiong (1997)
Tagetes minuta L.	Asteraceae	Salazar and Pignata (2014)
Thlaspi rotundifolium	Brassicaceae	Reeves and Brooks (1983)
Zea mays	Poaceae	Huang and Cunningham (1996)
Phaseolus vulgaris	Fabaceae	Luo et al. (2005)
Raphanus sativus	Brassicaceae	Chen et al. (2003)

metal from substratum, harvesting the plant produce and burning it for bio-ore (recovery of heavy metals) using specialist hyperaccumulator plant species (Ali et al. 2017; Ha et al. 2011). Phytomining gives precious metals, biofuel as well as increased soil nutrients and soil carbon contents (Brooks and Robinson 1998). Metal concentrations in the substratum and plant system, yearly productivity of plants, the biomass combustion energy and cost of recovered metal at international level influence the economics of phytomining (Brooks and Robinson 1998). Some biogeochemical factors viz. rhizobiological activity, exudates release, extended time, temperature, pH, damping of soil are some of the rate limiting factors for phytomining (Ali et al. 2013; Bhargava and Srivastava 2014). It is very difficult to remove lead once lead introduced into the soil matrix. Enhanced uptake of lead from the soil medium was observed at increased pH value, cation exchange capacity; soil/water Eh, content of organic carbon and phosphate levels (USEPA 1992). A model suggests that precipitation of lead as Pb-phoshate and effective roots mass are important factors for uptake and accumulation of lead into the plants (Brennan and Shelley 1999). Lead accumulation was highest in *Agrostemma githago* which is an herbaceous plant species, some of which produce enough biomass to be of practical use for phytoextraction of lead (Pichtel et al. 2000). *T. officinale* and *Ambrosia artemisiifolia* were reported as lead accumulator species in a study (Pichtel et al. 2000). Many members of families Brassicaceae, Euphorbiaceae, Asteraceae, Lamiaceae, and Scrophulariaceae were recognized as good accumulator of lead (Alkorta et al. 2004). Good amount of lead translocation from roots to the shoots is a well-known ability of *Brassica juncea* (Liu et al. 2000). In lead polluted soils *T. rotundifolium* has also been found to grow. Low metal bioavailability is the key reason limiting the potential of plant uptake for lead phytoextraction. Synthetic or natural chelators have been suggested to be mixed to the farm soil to trounce the said restraints (USEPA 2000a, 2000b). *Sesbania drummondii* accumulates up to 10,000 mg Pb kg^{-1} in aerial parts after exposure to a Pb-contaminated solution in hydroponic conditions (Sahi et al. 2002). Uptake of lead was found to increase by 21% after addition of EDTA (100 µM) to a medium containing 1 g Pb^{-1}. *Nicotiana glauca* R. Graham (shrub tobacco) a genetically modified variety has enormous ability to uptake lead, and found useful phytoremediation programmes (Gisbert et al. 2003). A protein (NtCBP4) that can alter plant tolerance to heavy metals was discovered by Arazi et al. (1999). For enhanced phytoremediation this gene may be valuable. Superior tolerance to nickel and hypersensitivity to lead, which are associated with inhibited nickel uptake and improved lead accumulation, respectively has been demonstrated in many separate transgenic lines expressing higher NtCBP4 gene (Arazi et al. 1999).

Phytostabilization (Phytoimmobilization)

In this technique plants reduce the mobility and migration of contaminants to soil, groundwater and food chain or stabilization the contaminants in contaminated soil through sorption, precipitation, accumulation and absorption by root (Erakhrumen 2007; Wuana and Okieimen 2011; Singh et al. 2012). Leachable constituents of contaminated environment make up a stable mass by absorption and binding around the plant system out of which the toxic pollutants cannot release in the surrounding environment. It is a management strategy only and cannot be a permanent solution for clean up contaminated sites (Vangronsveld et al. 2009). *Chrysopogon zizanioides* (vetiver grass) is an excellent option for phytostabilization, a method in which plants are used for the immobilization of pollutants in situ because it has ability to accumulate large concentrations of lead (Wilde et al. 2005) (Table 5). A small portion is transferred into the shoots while the majority of lead accumulated in the roots of vetiver grass. The solutions in the intercellular spaces in the roots have higher pH and comparatively higher levels of and carbonate-bicarbonates and phosphate; consequently accumulated lead is precipitated in the forms of phosphates/carbonates and prohibits translocation of lead in to the aerial parts (Danh et al. 2009, 2012). Extraordinary higher concentrations of lead are accumulated in the biomass of vetiver and can accumulate lead at least 1000 mg kg^{-1} DW. Vetiver can uptake over 10,000 and 3000 mg kg^{-1} Pb in roots and shoots, respectively, and accumulation of lead depends on the bioavailability of lead (Antiochia et al. 2007; Andra et al. 2009). Among many chelators, EDTA has been proved to be the most useful in the translocation of lead and a noteworthy increase of lead values in bio-mass of vetiver was observed when EDTA was applied in lead polluted medium (Danh et al. 2009, 2012).

Phytotransformation

Phytodegradation/phytotranformation refers to the mobilization and degradation of organic contaminants taken up by plants from soil and water and subsequently breaking down of pollutants at outside environment by various enzymes (dehaloge-nase and oxygenase) released by the plant systems. The characteristics of plants as well as the properties of the contaminants (solubility, hydrophobicity, polarity, etc.) affect the uptake of toxicants. Phytotransformation is independent from the activi-ties of microorganisms that present around root and in rhizosphere (Vishnoi and Srivastava 2008). The limitation of this technique is that it can be used for removal of heavy metal only, due to non-biodegradable nature of heavy metals. To short out this problem some synthetic herbicides, insecticides and transgenic plants are used by researchers recently (Doty 2008).

Phytofiltration

During this operation movement of toxic substances into underground waters is minimized through absorption or adsorption of contaminants. It is the elimination of contaminants from polluted water reservoirs or wastewaters using plant systems. On the basis of application of plant organs, phytofiltration has been classified as blast filtration when seedlings are in use; caulofiltration when plant shoots are in use and rhizofiltration when plant roots are in use (Ali et al. 2013). Contaminants in the soil solution adjacent the zone of roots are adsorbed or precipitated on roots or assimilation of these pollutants into the plant roots keep ongoing during the process of rhizofiltration. The plants to be made use for this intention are grown in green houses allowed to grow their roots rather in water in place of soil substratum. Once an outsized root system built up; from the polluted sites tainted water is collected and poured at these acclimatized plants for their water requirement. Root systems of plants growing in the contaminated region started to take up the contaminants along water. Saturated roots are used for the recovery of contaminants after harvesting and incinerated or composted (Singh et al. 2009; Pratas et al. 2012; Jagetiya et al. 2014)

Phytostimulation or Rhizodegradation

It is the breaking up of toxicants and pollutants in soil through microbes present in the rhizosphere. This phenomenon is also termed as plant-assisted bioremediation/ degradation or improved rhizosphere biodegradation (Mukhopadhyay and Maiti 2010) and always works at slow rates than phytodegradation. Plant roots secretes many natural biological compounds including sugars, alcohols, amino acids, and flavonoids. which provides nitrogen and carbon for rhizosphere microbes, and makes a nutrient affluent situation. Organic substances like solvents or petroleum fuel that is hazardous to living beings may be digested by various microbial species and they may breakdown these into nontoxic products through biodegradation. A large number of microbial species have been reported that have the ability to facilitate the oxidation of Fe^{2+} to Fe^{3+} (Jagetiya and Sharma 2009; Jagetiya et al. 2014).

Phytovolatilization

For removal of organic contaminants and volatile heavy metals such as Se and Hg, phytovolatilization is a preferred solution. Plants take up the contaminants from the environment and convert these into volatile form or a modified form with release into the atmosphere during transpiration. This process does not take away the contaminants thoroughly for that reason there are chances of re-deposition are always there (Ali et al. 2013).

5 Bioavailability of Lead

Bioavailability represents the amount of an element or compound available in soil system that is approachable to uptake by plant across its plasma membrane. Process in which plant absorb contaminants from soil through physiological membrane involve following four steps:

1. Solid-bound contaminant
2. Subsequent transport
3. Transport of bound contaminants (symplast/apoplast)
4. Uptake across a physiological membrane

Bioavailability of lead depends on physic-chemical properties of soil and activity of soil micro-and macro-organisms. Soil pH, ion exchange capacity, texture, porosity, age, adsorption capacity and environmental condition influence bioavailability of lead. Absorption efficiency or bioavailability of metal can be increased in soil by using some chelators consequently it facilitates the process of uptake of metals by plants. Stabilization of lead in contaminated soils can be achieved by adding phosphorus that reduces bioavailability (Chen et al. 2006). Lime and red mud also decrease lead availability to plants (Garau et al. 2007). Temperature also affect the bioavailability of lead, it is higher in warm than in cold environment (Hooda and Lloway 1993). Size and composition of lead particles affects the lead bioavailability to the plants (Walraven et al. 2015). Bone char addition in soil decreases the availability of lead. Free ionic form of lead (Pb^{2+}) is the largely bio-available and most toxic form which is present in the water whereas, chlorides, carbonates, and lead-organic matter complexes in fresh water or marine are other forms readily available to plants. Glomalin protein that is produced by AM fungi is binds mainly with lead in soil and reduces its bioavailability (Vodnik et al. 2008). Bio-surfactants (e.g. Di-rhamnolipid from *Pseudomonas aeruginosa*) or surfactants (e.g. DPC, DDAC, SDS and Gemini) facilitate the bioavailability process of lead in soil without effecting soil microorganisms and soil structure (Juwarkar et al. 2007; Mao et al. 2015). In some methods like sequential extraction, X-ray diffraction analysis, bioassay (Chen et al. 2006), and sorption processes, lead stabilization can be followed to examine the bioavailability of lead (Kumpiene et al. 2008). Bioavailability of lead can also be determined by bioluminescent bacterial reporter strains (Magrisso et al. 2009).

6 Lead: Uptake, Translocation and Accumulation

Uptake, translocation and accumulation of lead involve absorption of lead from soil into the plants and further transport into the xylem and phloem of plant systems (Dalvi and Bhalerao 2013). After accumulation of lead in roots the primary bulk flow of lead occurs into the xylem and the secondary bulk flow of lead occurs into the phloem (Marschner 1986; Mengel and Kirkby 1987). Uptake and transport of metal ion through root surface to vascular system is passive (pores of cell wall) or active (symplast). In this process metal ion and different special plasma membrane protein bind each other according to their analogous structure for transportation. Model plant *Arabidopsis thaliana* has 150 different cation transporter proteins. For example in *Thalspi rotundifolium* (lower biomass plant) can accumulate more lead in the roots than the shoots. In *Zea mays* (higher biomass plant) lead can move efficiently into shoots. To overcome this problem, soil amendments are performed (addition of chelators) to increase bioavailability of lead (Brennan and Shelley 1999). Heavy metals sequestration usually takes place in the vacuoles of the plant cells, where the metal/metal-ligand have to be brought across the tonoplast, the membrane of the vacuoles (Peer et al. 2005; Jagetiya and Sharma 2013).

7 Phytoremediation of Lead: Future Prospects

Phytoremediation is used for clean up toxic contaminants from environment with little environmental disturbance and good public perception. It has some limitations such as this process is very time consuming and toxic substances are accumulated in lower quantity which does not give large scale production in short time (Liu et al. 2000; Tangahu et al. 2011; Fukuda et al. 2014; Ali et al. 2017). To overcome this problem use of chelators that are biodegradable may enhances the process of phytoextraction as well as use of fast growing and hyperaccumulator-high biomass plants is recommended (Tandy et al. 2006; Evangelou et al. 2007). Advancement in molecular biology and genetic engineering can be make use to prepare genetically modify crops and transgenic plants that will helpful in further improvement in efficiency of phytoremediation (Tong et al. 2004; Ali et al. 2013). Many plant cultivars like

Cynodon dactylon, Vetiveria zizanioides, Festuca rubra and *Typha latifolia* are highly tolerant to temperature, flood, drought and toxic metals have been used recently. Vetiver grass (*Vetiveria zizanioides*) has reported to exhibit as a fine plant in phytoremediation of lead in china (Oh et al. 2014). In order to develop commercially and economically viable practices we need to optimize the agronomical systems, plant–microbe combinations in better way as well as plant genetic abilities (Jagetiya and Sharma 2009; Jagetiya et al. 2014). Genetic transformation of plant will help to overcome the limitation of this green technology through integrating some alien gene in plants for transporter proteins of metals, biosynthesis of enzymes required for sulphur metabolism (Kotrba et al. 2009). These modifications may enhance tolerance, uptake rate, detoxification capabilities of plants and biodegradation competence of microorganisms. Production of genetically modified plant can be successfully employed to promote some processes such as phytoextraction of metals (mainly Cd, Pb, Cu), breakdown of explosives and removal of carbon tetrachloride, vinyl chloride, benzene and chloroform (toxic volatile organic pollutants). These contaminants may be partially metabolized inside the plant tissues through "green liver" concept which involves three different steps, activation, conjugation and sequestration. A family of many enzymes normally involved in the metabolism of lethal and deadly contaminants has been recognized in *Populus angustifolia*. Enhanced heavy metal accumulation capability is proved in *Nicotiana tabacum* and *Silene cucubalus* (Fulekar et al. 2009). Advanced genetic strategies, use of transgenic plants and microbe will be able to contribute to the safer and wider applications of phytoremediation (Pence et al. 2000; Krämer and Chardonnens 2001; Ali et al. 2013; Jagetiya and Porwal 2019).

References

Abhilash PC, Jamil S, Singh N (2009) Transgenic plants for enhanced biodegradation and phytoremediation of organic xenobiotics. Biotechnol Adv 27:474–488

Addo MA, Darko EO, Gordon C, Nyarko BJB, Gbadago JK, Nyarko E, Affum HA, Botwe BO (2012) Evaluation of heavy metals contamination of soil and vegetable in the vicinity of a cement factory in the Volta Region, Ghana. Int J Sci Tech 2:40–50

Adesodun JK, Atayese MO, Agbaje TA, Osadiaye BA, Mafe OF, Soretire AA (2010) Phytoremediation potentials of sunflowers (*Tithonia diversifolia* and *Helianthus annuus*) for metals in soils contaminated with zinc and lead nitrates. Water Air Soil Pollut 207:195–201

Adriano DC (2001) Trace elements in terrestrial environments: biogeochemistry, bioavailability and risks of metals. Springer, New York, pp 223–232

Aery NC, Sarkar S, Jagetiya B, Jain GS (1994) Cadmium-zinc tolerance in soybean and fenugreek. J Ecotoxicol Environ Monit 4:39–44

Aery NC, Jagetiya B (1997) Relative toxicity of cadmium, lead and zinc on barley. Commun Soil Sci Plant Anal 28:949–960

Agency for Toxic Substances and Disease Registry (ATSDR) (2005) Toxicological profile for lead. (Draft for Public Comment). Atlanta, GA: U.S. Department of Health and Human Services, Public Health Service, pp 43–59.

Ahamed M, Siddiqui MKJ (2007) Environmental lead toxicity and nutritional factors. Clin Nutr 26:400–408

Ahamed M, Verma S, Kumar A, Siddiqui MK (2005) Environmental exposure to lead and its correlation with biochemical indices in children. Sci Total Environ 346:48–55

Ajavan KV, Selvaraju M, Thirugnanamoorthy K (2011) Growth and heavy metals accumulation potential of microalgae grown in sewage wastewater and petrochemical effluents. Pak J Biol Sci 14:805–811

Akpor OB, Muchie M (2011) Environmental and public health implications of wastewater quality. Afr J Biotechnol 10:2379–2387

Ali A, Guo D, Mahara A, Ping W, Wahid F, Shen F, Li R, Zhang Z (2017) Phytoextraction and the economic perspective of phytomining of heavy metals. Solid Earth 75:1–40

Ali EH, Hashem M (2007) Removal efficiency of the heavy metals Zn(II), Pb(II) and Cd(II) by *Saprolegnia delica* and *Trichoderma viride* at different pH values and temperature degrees. Mycobiology 35:135–144

Ali H, Naseer M, Sajad MA (2012) Phytoremediation of heavy metals by *Trifolium alexandrinum*. Int J Env Sci 2:1459–1469

Ali H, Khan E, Sajad MA (2013) Phytoremediation of heavy metals-concepts and applications. Chemosphere 91:869–881

Alkorta I, Hernández-Allica J, Becerril JM, Amezaga I, Albizu I, Garbisu C (2004) Recent findings on the phytoremediation of soils contaminated with environmentally toxic heavy metals and metalloids such as zinc, cadmium, lead, and arsenic. Rev Environ Sci Biotechnol 3:71–90

Al-Masri MS, Mukalallati H, Al-Hamwi A (2014) Transfer factors of ^{226}Ra, ^{210}Pb and ^{210}Po from NORM-contaminated oilfield soil to some *Atriplex* species, Alfalfa and Bermuda grass. Radioprotection 49:27–33

Andra SS, Datta R, Sarkar D, Makris KC, Mullens CP, Sahi SV, Bach SB (2009) Induction of lead-binding phytochelatins in vetiver grass (*Vetiveria zizanioides* L.). J Environ Qual 38:868–877

Antiochia R, Campanella L, Ghezzi P, Movassaghi K (2007) The use of vetiver for remediation of heavy metal soil contamination. Anal Bioanal Chem 388:947–956

Arazi T, Sunkar R, Kaplan B, Fromm H (1999) A tobacco plasma membrane calmodulin-binding transporter confers Ni^{2+} tolerance and Pb^{2+} hypersensitivity in transgenic plants. Plant J 20:171–182

Arora K, Sharma S, Monti A (2015) Bio-remediation of Pb and Cd polluted soils by switchgrass: a case study in India. Int J Phytoremediation 17:285–321

Athalye VV, Mistry KB (1972) Uptake and distribution of polonium-210 and lead-210 in tobacco plants. Radiat Bot 12:421–425

Aung WL, Aye KN, Hlaing NN (2012) Biosorption of lead (Pb^{2+}) by using *Chlorella vul- garis*. In: Proceedings of International Conference on Chemical Engineering and its Applications. Bangkok, Thailand

Azizi SN, Colagar AH, Hafeziyan SM (2012) Removal of Cd (II) from aquatic system using *Oscillatoria* sp. biosorbent. Scient World J 2012:347053

Baker AJM, Reeves RD, McGrath SP (1991) In situ decontamination of heavy metal polluted soils using crops of metal-accumulating plants-A feasibility study. In: Hinchee RE, Olfenbuttel RF (eds) In situ bioreclamation: applications and investigations for hydrocarbon and contaminated sites remediation. Butterworth-Heinemann, London, pp 600–605

Balsberg-Påhlsson AM (1989) Toxicity of heavy metals (Zn, Cu, cd, Pb) to vascular plants. A literature review. Water Air Soil Pollut 47:287–319

Basha SA, Rajaganesh K (2014) Microbial bioremediation of heavy metals from textile industry dye effluents using isolated bacterial strains. Int J Curr Microbiol App Sci 3:785–794

Bech J, Duran P, Roca N, Poma W, Sánchez I, Barceló J, Boluda R, Roca-Pérez L, Poschenrieder C (2012) Shoot accumulation of several trace elements in native plant species from contaminated soils in the Peruvian Andes. J Geochem Explor 113:106–111

Beolchini F (2006) Ionic strength effect on copper biosorption by *Sphaerotilus natans*: equilibrium study and dynamic modeling in membrane reactor. Water Res 40:144–152

Bhargava A, Srivastava S (2014) Transgenic approaches for phytoextraction of heavy metals. In: Ahmad P, Wani MR, Azooz MM, Tran LSP (eds) Improvement of crops in the era of climatic changes. Springer, New York, pp 57–80

Bisessar S (1981) Effect of heavy metals on microorganisms in soils near a secondary lead smelter. Water Air Soil Pollut 17:305–308

Brennan MA, Shelley ML (1999) A model of the uptake, translocation, and accumulation of lead (Pb) by maize for the purpose of phytoextraction. Ecol Eng 12:271–297

Brooks RR, Robinson BH (1998) The potential use of hyperaccumulators and other plants in phytomining. In: Brooks RR (ed) Plants that hyperaccumulate heavy metals: their role in phytoremediation, microbiology, archaeology, mineral exploration and phytomining. CAB International, Wallingford, UK, pp 327–356

Chakraborty P, Babu PVR, Sarma VV (2012) A study of lead and cadmium speciation in some estuarine and coastal sediment. Chem Geol 294-295:217–225

Chakraborty S, Chakraborty P, Nath BN (2015) Lead distribution in coastal and estuarine sediments around India. Mar Pollut Bull 97:36–46

Chandrasekhar C, Ray JG (2019) Lead accumulation, growth responses and biochemical changes of three plant species exposed to soil amended with different concentrations of lead nitrate. Ecotoxicol Environ Saf 171:26–36

Chapman PM (2002) Integrating toxicology and ecology: putting the "eco" into ecotoxicology. Mar Pollut Bull 44:7–15

Chehregani A, Malayeri BE (2007) Removal of heavy metals by native accumulator plants. Int J Agric Bio 9:462–465

Chehregani A, Noori M, Yazdi HL (2009) Phytoremediation of heavy-metal-polluted soils: Screening for new accumulator plants in Angouran mine (Iran) and evaluation of removal ability. Ecotoxicol Environ Saf 72:1349–1353

Chen YX, Lin Q, Luo YM, He YF, Zhen SJ, Yu YL, Tian GM, Wong MH (2003) The role of citric acid on the phytoremediation of heavy metal contaminated soil. Chemosphere 50(6):807–811

Chen Y, Li X, Shen Z (2004) Leaching and uptake of heavy metals by ten different species of plants during an EDTA-assisted phytoextraction process. Chemosphere 57:187–196

Chen SB, Zhu YG, Ma YB, McKay G (2006) Effect of bone char application on Pb bioavailability in a Pb-contaminated soil. Environ Pollut 139:433–439

Choi SB, Yun YS (2004) Lead bio-sorption by waste biomass of *Corynebacterium glutamicum* generated from lysine fermentation process. Biotechnol Lett 26:331–336

Chou FI, Chung HP, Teng SP, Sheu ST (2005) Screening plant species native to Taiwan for remediation of ^{137}Cs-contaminated soil and the effects of K addition and soil amendment on the transfer of ^{137}Cs from soil to plants. J Environ Radioact 80:175–181

Cohen AJ, Roe FJC (1991) Review of lead toxicology relevant to the safety assessment of lead acetate as a hair colouring. Food Chem Toxicol 29:485–507

Corneanu M, Gabriel CC, Crăciun C, Tripon S (2014) Phytoremediation of some heavy metals and radionuclides from a polluted area located on the middle Jiu river. Case study: *Typha latifolia* L. *Muzeul Olteniei* Craiova. Oltenia Studii Şi Comunicǎri Ştiinţele Naturii Tom 30:1454–6924

D'Souza TJ, Mistry KB (1970) Comparative uptake of thorium-230, radium-226, lead-210 and polonium-210 by plants. Radiat Bot 10:293–295

Dalvi AA, Bhalerao SA (2013) Response of plants towards heavy metal toxicity: an overview of avoidance, tolerance and uptake mechanism. Ann Plant Sci 2:3262–3268

Damodaran D, Suresh G, Mohan RB (2011) Bioremediation of soil by removing heavy metals using *Saccharomyces cerevisiae*. IACSIT Press, Singapore

Danh LT, Truong P, Mammucari R, Tran T, Foster N (2009) Vetiver grass, *Vetiveria zizanioides*: a choice plant for phytoremediation of heavy metals and organic wastes. Int J Phytorem 11:664–691

Danh LT, Truong P, Mammucari R, Pu Y, Foster NR (2012) Phytoremediation of soils contaminated by heavy metals, metalloids, and radioactive materials using vetiver grass, *Chrysopogon zizanioides*. In: Anjum NA, Pereira ME, Ahmad I, Duarte AC, Umar S, Khan N (eds) Phytotechnologies: remediation of environmental contamination. CRC Press, pp 278–303

Degraeve N (1981) Carcinogenic, teratogenic and mutagenic effects of cadmium. Mutat Res 86:115–135

Deng L, Su Y, Su H, Wang X, Zhu X (2007) Sorption and desorption of lead (II) from wastewater by green algae *Cladophora fascicularis*. J Hazard Mater 143:220–225

Dixit R (2015) Bioremediation of heavy metals from soil and aquatic environment: an overview of principles and criteria of fundamental processes. Sustainability 7:2189–2212

Djingova R, Kuleff I (2000) Instrumental techniques for trace analysis. In: Markert B, Friese K (eds) Trace elements: their distribution and effects in the environment. Elsevier, pp 137–185

Dogan M, Karatasb M, Aasim M (2018) Cadmium and lead bioaccumulation potentials of an aquatic macrophyte *Ceratophyllum demersum* L.: A laboratory study. Ecotoxicol Environ Saf 148:431–440

Doty SL (2008) Enhancing phytoremediation through the use of transgenics and entophytes. New Phytol 179:318–333

Dwivedi S (2012) Bioremediation of heavy metal by algae: current and future perspective. J Adv Lab Res Biol:195–199

Dwivedi S, Mishra A, Saini D (2012) Removal of heavy metals in liquid media through fungi isolated from wastewater. Int J Sci Res 1:181–185

Eapen S, Singh S, Thorat V, Kaushik CP, Raj K, D'Souza SF (2006) Phytoremediation of radio-strontium (^{90}Sr) and radiocesium (^{137}Cs) using giant milky weed (*Calotropis gigantea* R. Br.) plants. Chemosphere 65:2071–2073

Edris G, Alhamed Y, Alzahrani A (2012) Cadmium and lead biosorption by *Chlorella vulgaris*. In: IWTA, 16th International Water Technical Conference. Istanbul, Turkey

Epelde L, Mijangos I, Becerril JM, Garbisu C (2009) Soil microbial community as bioindicator of the recovery of soil functioning derived from metal phytoextraction with sorghum. Soil Biol Biochem 41:1788–1794

Erakhrumen AA (2007) Phytoremediation: an environmentally sound technology for pollution prevention, control and remediation in developing countries. Educ Res Rev 2:151–156

Evangelou MWH, Ebel M, Schaeffer A (2007) Chelate assisted phytoextraction of heavy metals from soil: effect, mechanism, toxicity and fate of chelating agents. Chemosphere 68:989–1003

Fanna AG, Yadji G, Abdourahmane TDB, Zakaria OI, Karimou AJM (2018) Phytoextraction of Pb, Cd, Cu and Zn by *Ricinus communis*. Environ Wat Sci Pub Health Ter Int J 2:56–62

Flora G, Gupta D, Tiwari A (2012) Toxicity of lead: a review with recent updates. Interdis Toxicol 5:47–58

Fu F, Wang Q (2011) Removal of heavy metal ions from wastewaters: a review. J Environ Manage 92:407–418

Fukuda S, Iwamoto K, Atsumi M, Yokoyama A, Nakayama T, Ishida KI, Inouhe I, Shiraiwa Y (2014) Current status and future control of cesium contamination in plants and algae in Fukushima Global searches for microalgae and aquatic plants that can eliminate radioactive cesium, iodine and strontium from the radio-polluted aquatic environment: a bioremediation strategy. J Plant Res 127:79–89

Fulekar MH, Singh A, Bhaduri AM (2009) Genetic engineering strategies for enhancing phytoremediation of heavy metals. Afr J Biotechnol 8:529–535

Fulghum JE, Bryant SR, Linton RW, Grlffls DP (1988) Discrimination between adsorption and coprecipitation in aquatic particle standards by surface analysis techniques: lead distributions in calcium carbonates. Environ Sci Technol 22:463–467

Garau G, Castaldi P, Santona L, Deiana P, Melis P (2007) Influence of red mud, zeolite and lime on heavy metal immobilization, culturable heterotrophic microbial populations and enzyme activities in a contaminated soil. Geoderma 142:47–57

Garbisu C, Alkorta I (2003) Basic concepts on heavy metal soil bioremediation. Eur J Miner Process Environ Prot 3:58–66

Ghrefat H, Yusuf N (2006) Assessing Mn, Fe, Cu, Zn and Cd pollution in bottom sediments of Wadi Al-Arab Dam, Jordan. Chemosphere 65:2114–2121

Gisbert C, Ros R, de Haro A, Walker DJ, Bernal MP, Serrano R, Navarro-Avino J (2003) A plant genetically modified that accumulates Pb is especially promising for phytoremediation. Biochem Biophys Res Commun 303:440–445

Guilizzoni P (1991) The role of heavy metals and toxic materials in the physiological ecology of submersed macrophytes. Aquat Bot 41:87–109

Gulati K, Banerjee B, Lall SB, Ray A (2010) Effects of diesel exhaust, heavy metals and pesticides on various organ systems: possible mechanisms and strategies for prevention and treatment. Indian J Exp Biol 48:710–721

Gupta DK, Nicoloso FT, Schetinger MR, Rossato LV, Pereira LB, Castro GY, Srivastava S, Tripathi RD (2009) Antioxidant defense mechanism in hydroponically grown *Zea mays* seedlings under moderate lead stress. J Hazard Mater 172:479–484

Gupta DK, Huang HG, Yang XE, Razafindrabe BH, Inouhe M (2010) The detoxification of lead in *Sedum alfredii* H. is not related to phytochelatins but the glutathione. J Hazard Mater 177:437–444

Gupta DK, Huang HG, Corpas FJ (2013a) Lead tolerance in plants: strategies for phytoremediation. Environ Sci Pollut Res 20:2150–2161

Gupta DK, Huang HG, Nicoloso FT, Schetinger MR, Farias JG, Li TQ, Razafindrabe BH, Aryal N, Inouhe M (2013b) Effect of Hg, As and Pb on biomass production, photosynthetic rate, nutrients uptake and phytochelatin induction in *Pfaffia glomerata*. Ecotoxicology 22:1403–1412

Ha NT, Sakakibara M, Sano S (2011) Accumulation of Indium and other heavy metals by *Eleocharis acicularis*: an option for phytoremediation and phytomining. Bioresour Technol 102:2228–2234

Ham GJ, Wilkins BT, Ewers LW (2001) ^{210}Pb, ^{210}Po, ^{226}Ra, U and Th in arable crops and ovine liver: variations in concentrations in the United Kingdom and resultant doses. Radiat Prot Dosimetry 93:151–159

He LY, Chen ZJ, Ren GD, Zhang YF, Qian M, Sheng XF (2009) Increased cadmium and lead uptake of a cadmium hyperaccumulator tomato by cadmium-resistant bacteria. Ecotoxicol Environ Saf 72:1343–1348

Holan ZR, Volesky B (1994) Biosorption of lead and nickel by biomass of marine algae. Biotechnol Bioeng 43:1001–1009

Hooda PS, Lloway BLA (1993) Effects of time and temperature on the bioavailability of Cd and Pb from sludge-amended soils. J Soil Sci 44:97–110

Hovmand MF, Nielsen SP, Johnsen I (2009) Root uptake of lead by Norway spruce grown on 210Pb spiked soils. Environ Pollut 157:404–409

Hseu ZY, Jien SH, Wang SH, Deng HW (2013) Using EDDS and NTA for enhanced phytoextraction of Cd by water spinach. J Environ Manage 117:58–64

Huang H, Gupta DK, Tian S, Yang XE, Li T (2011) Lead tolerance and physiological adaptation mechanism in roots of accumulating and non-accumulating ecotypes of *Sedum alfredii*. Environ Sci Pollut Res 19:1640–1651

Huang JW, Cunningham SD (1996) Lead phytoextraction: species variation in lead uptake and translocation. New Phytol 13:75–84

Inouhe M, Sakuma Y, Chatterjee S, Datta S, Jagetiya B, Voronina AV, Walther C, Gupta DK (2015) General roles of phytochelatins and other peptides in plant defense mechanisms against oxidative stress/primary and secondary damages induced by heavy metals. In: Gupta DK, Palma JM, Corpas FJ (eds) Reactive oxygen species and oxidative damage in plants. Springer, Cham, pp 1–22

Jagetiya B, Aery NC (1994) Effect of low and toxic levels of nickel on seed germination and early seedling growth of moong. Bionature 14:57–61

Jagetiya B, Bhatt K (2005) Nickel induced biochemical and physiological alterations in barley. Bionature 25:75–81

Jagetiya B, Purohit P (2006) Effect of different uranium tailing concentrations on certain growth and biochemical parameters in sunflower. Biologia 61:103–107

Jagetiya B, Bhatt K (2007) Relative toxicity of various nickel species on seed germination and early seedling growth of *Vigna unguiculata* L. Asian J Bio Sci 2:11–17

Jagetiya B, Sharma A (2009) Phytoremediation of radioactive pollution: present status and future. Ind J Bot Res 5:45–78

Jagetiya B, Sharma A (2013) Optimization of chelators to enhance uranium uptake from tailings for phytoremediation. Chemosphere 91:692–696

Jagetiya B, Porwal SR (2019) Exploration of floral diversity of polluted habitats around Bhilwara city for phytoremediation. Plant Arch 19:403–406

Jagetiya B, Bhatt K, Kaur MJ (2007) Activity of certain enzymes and growth as affected by Nickel. Ind J Bot Res 3:103–114

Jagetiya B, Soni A, Kothari S, Khatik U (2011) Bioremediation: an ecological solution to textile effluents. Asian J Bio Sci 6:248–257

Jagetiya B, Purohit P, Kothari S, Pareek P (2012) Influence of various concentrations of uranium mining waste on certain growth and biochemical parameters in gram. Int J Plant Sci 7:79–84

Jagetiya B, Soni A, Yadav S (2013) Effect of nickel on plant water relations and growth in green gram. Indian J Plant Physiol 18:372–376

Jagetiya B, Sharma A, Soni A and Khatik UK (2014) Phytoremediation of radionuclides: A report on the state of the art. In: Gupta DK, Walther C (eds) Radionuclide contamination and remediation through plants. Springer, Champ, pp 1–31

Jamil S, Abhilash PC, Singh N, Sharma PN (2009) *Jatropha curcas*: a potential crop for phytoremediation of coal fly-ash. J Hazard Mater 172:269–275

Johnson CE, Siccama TG, Driscoll CT, Likens GE, Moeller RE (1995) Changes in lead biogeochemistry in response to decreasing atmospheric inputs. Ecol Appl 5:813–822

Jones B, Turki A (1997) Distribution and speciation of heavy metals in surficial sediments from the Tees estuary, North-east England. Mar Pollut Bull 34:768–779

Joshi PK, Swarup A, Maheshwari S, Kumar R, Singh N (2011) Bioremediation of heavy metals in liquid media through fungi isolated from contaminated sources. Ind J Microbiol 51:482–487

Juberg DR, Kleiman CF, Kwon SC (1997) Position paper of the American council on science and health: lead and human health. Ecotoxicol Environ Saf 38:162–180

Juwarkar AA, Nair A, Dubey KV, Singh SK, Devotta S (2007) Biosurfactant technology for remediation of cadmium and lead contaminated soils. Chemosphere 68:1996–2002

Kapoor A, Viraraghavan T, Cullimore DR (1999) Removal of heavy metals using fungus *Aspergillus niger*. Bioresour Technol 70:95–104

Kapourchal SA, Kapourchal SA, Pazira E, Homaee M (2009) Assessing radish (*Raphanus sativus* L.) potential for phytoremediation of lead-polluted soils resulting from air pollution. Plant Soil Environ 55:202–206

Kotrba P, Najmanova J, Macek T (2009) Genetically modified plants in phytoremediation of heavy metal and metalloid soil and sediment pollution. Biotechnol Adv 27:799–810

Krämer U, Chardonnens AN (2001) The use of transgenic plants in bioremediation of soils contaminated with trace elements. Appl Microbiol Biotechnol 55:661–672

Krupadam RJ, Ahuja R, Wate SR (2007) Heavy metal binding fractions in the sediments of the Godavari estuary, East coast of India. Environ Model Assess 12:145–155

Kumar JI, Oommen C (2012) Removal of heavy metals by bio-sorption using fresh water alga *Spirogyra hyalina*. J Environ Biol 33:27–31

Kumar PBAN, Dushenkov V, Motto H, Raskin I (2002) Phytoextraction: The use of plants to remove heavy metals from soils. Environ Sci Technol 29:1232–1238

Kumpiene J, Lagerkvist A, Maurice C (2008) Stabilization of As, Cr, Cu, Pb and Zn in soil using amendments—a review. Waste Manag 28:215–225

Kusell M, Lake L, Andersson M, Gerschenson LE (1978) Cellular and molecular toxicology of lead. II. Effect of lead on δ-aminolevulinic acid synthetase of cultured cells. J Toxicol Environ Health 4:515–525

Lacerda LD (1998) Trace metals biogeochemistry and diffuse pollution in mangrove ecosystems. ISME Mangrove Ecosystems Occasional Papers 2:149–157

Lajayer BA, Ghorbanpour M, Nikabadi S (2017) Heavy metals in a contaminated environment: destiny of secondary metabolite biosynthesis, oxidative status and phytoextraction in medicinal plants. Ecotoxl Environ Saf 145:377–390

Li D, Tan XY, Wu XD, Pan C, Xu P (2014) Effects of electrolyte characteristics on soil conductivity and current in electrokinetic remediation of lead-contaminated soil. Sep Purif Technol 135:14–21

Lin CC, Lai YT (2006) Adsorption and recovery of lead (II) from aqueous solutions by immobilized *Pseudomonas aeruginosa* PU21 beads. J Hazard Mater 137:99–105

Liu D, Jiang W, Liu C, Xin C, Hou W (2000) Uptake and accumulation of lead by roots, hypocotyls and shoots of Indian mustard (*Brassica juncea* L.). Bioresour Technol 71:273–277

Luo C, Shen Z, Li X (2005) Enhanced phytoextraction of Cu, Pb, Zn and Cd with EDTA and EDDS. Chemosphere 59:1–11

Magrisso S, Belkin S, Erel Y (2009) Lead bioavailability in soil and soil components. Water Air Soil Pollut 202:315–323

Mahar A, Wang P, Ali A, Awasthi MK, Lahori AH, Wang Q, Li R, Zhang Z (2016) Challenges and opportunities in the phytoremediation of heavy metals contaminated soils: a review. Ecotoxicol Environ Saf 126:111–121

Malar S, Vikram SS, JC Favas P, Perumal V (2014) Lead heavy metal toxicity induced changes on growth and antioxidative enzymes level in water hyacinths [*Eichhornia crassipes* (Mart.)]. Bot Stud 55:54

Mamboya FA, Pratap HB, Mtolera M, Bjork M (1999) The effect of copper on the daily growth rate and photosynthetic efficiency of the brown macro alga *Padina boergensenii*. In: Richmond MD, Francis J (eds) Proceedings of the Conference on Advances on Marine Sciences in Tanzania. pp 185–192.

Mao X, Jiang R, Xiao W, Yu J (2015) Use of surfactants for the remediation of contaminated soils: a review. J Hazard Mater 285:419–435

Marschner P (1986) Mineral nutrition of higher plants. Academic Press, Orlando, FL

Massaccesi G, Romero MC, Cazau MC, Bucsinszky AM (2002) Cadmium removal capacities of filamentous soil fungi isolated from industrially polluted sediments, La Plata (Argentina). World J Microbiol Biotechnol 18:817–820

Meers E, Slycken SV, Adriaensen K, Ruttens A (2010) The use of bio-energy crops (*Zea mays*) for 'phytoattenuation' of heavy metals on moderately contaminated soils: a field experiment. Chemosphere 78:35–41

Mengel K, Kirkby EA (1987) Principles of plant nutrition. Springer, Dordrecht

Mitchell N, Pérez-Sánchez D, Thorne MC (2013) A review of the behaviour of U-238 series radionuclides in soils and plants. J Radiol Prot 33:17–48

Mudgal V, Madaan N, Mudgal A (2010) Heavy metals in plants: phytoremediation: plants used to remediate heavy metal poll. Agric Bio J North Amer 1:40–46

Mukhopadhyay S, Maiti SK (2010) Phytoremediation of metal enriched mine waste: a review. Glob J Environ Res 4:135–150

Nan Z, Cheng G (2001) Accumulation of Cd and Pb in spring wheat (*Triticum aestivum* L.) grown in calcareous soil irrigated with wastewater. Bull Environ Contam Toxicol 66:748–754

Neustadt J, Pieczenik S (2007) Toxic-metal contamination: Mercury. Integr Med 6:36–37

Oh K, Cao T, Li T, Cheng HY (2014) Study on application of phytoremediation technology in management and remediation of contaminated soils. J Clean Ener Technol 2:216–220

Okkenhaug G, Grasshorn Gebhardt KA, Amstaetter K, Lassen Bue H, Herzel H, Mariussen E, Mulder J (2016) Antimony (Sb) and lead (Pb) in contaminated shooting range soils: Sb and Pb mobility and immobilization by iron based sorbents, a field study. J Hazard Mater 307:336–343

Padmavathiamma PK, Li LY (2007) Phytoremediation technology: Hyper-accumulation metals in plants. Water Air Soil Pollut 184:105–126

Pan TL, Wang PW, Al Suwayeh SA, Chen CC, Fang JY (2010) Skin toxicology of lead species evaluated by their permeability and proteomic profiles: a comparison of organic and inorganic lead. Toxicol Lett 197:19–28

Patel KS, Shrivas K, Hoffmann P, Jakubowski N (2006) A survey of lead pollution in Chhattisgarh State, central India. Environ Geochem Health 28:11–17

Patil AJ, Bhagwat VR, Patil JA, Dongre NN, Ambekar JG, Jailkhani R, Das KK, Sheng PX (2006) Effect of lead (Pb) exposure on the activity of superoxide dismutase and catalase in battery manufacturing workers (BMW) of western maharashtra (India) with reference to heme biosynthesis. Int J Environ Res Public Health 3:329–337

Peer WA, Baxter IR, Richards EL, Freeman JL, Murphy AS (2005) Phytoremediation and hyperaccumulator plants. Topic Curr Genet 14:84

Pence NS, Larsen PB, Ebbs SD (2000) The molecular physiology of heavy metal transport in the Zn/Cd hyperaccumulator *Thlaspi caerulescens*. Proc Natl Acad Sci U S A 97:4956–4960

Pichtel J, Kuroiwa K, Sawyerr HT (2000) Distribution of Pb, Cd and Ba in soils and plants of two contaminated soils. Environ Pollut 110: 171-178.

Pietrzak-Fils Z, Skowronska-Smolak M (1995) Transfer of ^{210}Pb and ^{210}Po to plants via root system and above-ground interception. Sci Total Environ 162:139–147

Piotrowska NA, Bajguz A, Talarek M, Bralska M, Zambrzycka E (2015) The effect of lead on the growth, content of primary metabolites, and antioxidant response of green alga *Acutodesmus obliquus* (Chlorophyceae). Environ Sci Pollut Res 22:19112–19123

Prasad MNV, Freitas HM (2003) Metal hyperaccumulation in plants-biodiversity prospecting for phytoremediation technology. Electron J Biotechnol 6:285–321

Pratas J, Favas PJC, Paulo C, Rodrigues N, Prasad MN (2012) Uranium accumulation by aquatic plants from uranium-contaminated water in central Portugal. Int J Phytorem 14:221–234

Purvis OW, Halls C (1996) A review of lichens in metal-enriched environment. Lichenologist 28:571–601

Rajapaksha BE (2004) Metal toxicity affects fungal and bacterial activities in soil differently. Appl Environ Microbiol 70:2966–2973

Rana L, Chhikara S, Dhankar R (2013) Assessment of growth rate of indigenous cyanobacteria in metal enriched culture medium. Asian J Exp Bio 4:465–471

Reeves RD, Brooks RR (1983) Hyperaccumulation of lead and zinc by two metallophytes from mining areas of Central Europe. Environ Pollut 31:277–285

Reuer MK, Weiss DJ (2002) Anthropogenic lead dynamics in the terrestrial and marine environment. Phil Trans Royal Soc A 360:2889–2904

Rocchetta I, Leonardi PI, Amado Filho GM, Molina MDR, Conforti V (2007) Ultrastructure and x-raymicroanalysis of *Euglena gracilis* (Euglenophyta) under chromium stress. Phycologia 46:300–306

Rossi N, Jamet JL (2008) In situ heavy metals (copper, lead and cadmium) in different plankton compartments and suspended particulate matter in two coupled Mediterranean coastal ecosystems (Toulon Bay, France). Mar Pollut Bull 56:1862–1870

Ruttens A, Boulet J, Weyens N, Smeets K (2011) Short rotation coppice culture of willows and poplars as energy crops on metal contaminated agriculture soils. Int J Phytorem 13:194–207

Sadik R, Lahkale R, Hssaine N, ElHatimi W, Diouri M, Sabbar E (2015) Sulfate removal from wastewater by mixed oxide-LDH: equilibrium, kinetic and thermodynamic studies. J Mater Environ Sci 6:2895–2905

Sahi SV, Bryant NL, Sharma NC, Singh SR (2002) Characterization of a lead hyperaccumulator shrub, *Sesbania drummondii*. Environ Sci Technol 36:4676–4680

Sakakibara M, Ohmori Y, Ha NTH, Sano S, Sera K (2011) Phytoremediation of heavy metal-contaminated water and sediment by *Eleocharis acicularis*. CLEAN-Soil Air Water 39:735–741

Salazar MJ, Pignata ML (2014) Lead accumulation in plants grown in polluted soils. Screening of native species for phytoremediation. J Geochem Explor 137:29–36

Salehizadeh H, Shojaosadati SA (2003) Removal of metal ions from aqueous solution by polysaccharide produced from *Bacillus firmus*. Water Res 37:4231–4235

Salem HM, Eweida EA, Farag A (2000) Heavy metals in drinking water and their environmental impact on human health. In: The proceedings of ICEHM 2000 meeting, Cairo University, Egypt, pp 542–556

Santos RWD, Schmidt ÉC, Felix MRD, Polo LK, Kreusch M, Pereira DT, Costa GB, Simioni C, Chow F, Ramlov F, Maraschin M, Bouzon ZL (2014) Bioabsorption of cadmium, copper and lead by thread macro alga *Gelidium floridanum*: Physiological responses and ultrastructure features. Ecotoxicol Environ Saf 105:80–89

Schreck E, Foucault Y, Geret F, Pradere P, Dumat C (2011) Influence of soil ageing on bioavailability and ecotoxicity of lead carried by process waste metallic ultrafine particles. Chemosphere 85:1555–1562

Sekhar KC, Kamala CT, Chary NS, Balaram V, Garcia G (2005) Potential of *Hemidesmus indicus* for phytoextraction of lead from industrially contaminated soils. Chemosphere 58:507–514

Sharma P, Dubey RS (2005) Lead toxicity in plants. Braz J Plant Physiol 17:35–52

Sheng PX, Ting Y, Chen JP, Hong L (2004) Sorption of lead, copper, cadmium, zinc and nickel by marine algal biomass: characterization of biosorptive capacity and investigation of mechanisms. J Colloid Interface Sci 275:131–141

Sheoran V, Sheoran A, Poonia P (2011) Role of hyperaccumulators in phytoextraction of metals from contaminated mining sites: a review. Crit Rev Environ Sci Technol 41:168–214

Sheppard SC, Sheppard MI, Sanipelli BL, Tait JC (2004) Background radionuclide concentrations in major environmental compartments of natural ecosystems. Report by Eco Matters for the Canadian Nuclear Safety Commission Contract No. 87055020215

Sheppard SC, Sheppard MI, Ilin M, Tait J, Sanipelli B (2008) Primordial radionuclides in Canadian background sites: secular equilibrium and isotopic differences. J Environ Radioact 99:933–946

Shoty K, Weiss DW, Appleby PG, Chebrkin AK, Gloor RFM, Kramens JD (1998) History of atmosphearic lead deposition since 12,370 (14)C yr BP from a peat bog, Jura Mountains, Switzerland. Science 281:1635–1640

Singh D, Tiwari A, Gupta R (2012) Phytoremediation of lead from wastewater using aquatic plants. J Agr Technol 8:1–11

Singh D, Vyas P, Sahni S, Sangwan P (2015) Phytoremediation: a biotechnological intervention. In: Kaushik G (ed) Applied environmental biotechnology: present scenario and future trends. Springer, New Delhi, pp 59–75

Singh S (2012) Phytoremediation: a sustainable alternative for environmental challenges. Int J Gr Herb Chem 1:133–139

Singh S, Thorat V, Kaushik CP, Raj K, Eapan S, D'Souza SF (2009) Potential of Chromolaena odorata for phytoremediation of ^{137}Cs from solution and low level nuclear waste. J Hazard Mater 162:743–745

Singh VK, Mishra KP, Rani R, Yadav VS, Awasthi SK, Garg SK (2003) Immunomodulation by lead. Immunol Res 28:151–166

Sprang PAV, Nys C, Blust RJP, Chowdhury J, Gustafsson JP, Janssen CJ, Schamphelaere KACD (2016) The derivation of effects threshold concentrations of lead for European freshwater ecosystems. Environ Toxicol Chem 35:1310–1320

Sugihara S, Efrizal Osaki S, Momoshima N, Maeda Y (2008) Seasonal variation of natural radionuclides and some elements in plant leaves. J Radioanal Nucl Chem 278:419–422

Tandy S, Schulin R, Nowack B (2006) The influence of EDDS in the uptake of heavy metals in hydroponically grown sunflowers. Chemosphere 62:1454–1463

Tang YT, Qiu RL, Zeng XW, Ying RR, Yu FM, Zhou XY (2009) Lead, zinc, cadmium hyperaccumulation and growth stimulation in *Arabis paniculata* Franch. Environ Exp Bot 66:126–134

Tangahu V, Abdullah SRS, Basri H, Idris M, Anuar N, Mukhlisin M (2011) A review on heavy metals (As, Pb, and Hg) uptake by plants through phytoremediation. Int J Chem Eng 2011:939161

Tiwari S, Tripathi IP, Tiwari H (2013) Blood lead level—a review. Int J Eng Sci Technol 3:330–333

Tong YP, Kneer R, Zhu YG (2004) Vacuolar compartmentalization: a second-generation approach to engineering plants for phytoremediation. Trend Plant Sci 9:7–9

Truhaut R (1977) Eco-toxicology-objectives, principles and perspectives. Ecotoxicol Environ Saf 1:151–173

USEPA (1992) Selection of control technologies for remediation of lead battery recycling sites. EPA/540/S-92/011. US Environmental Protection Agency, Office of Emergency and Remedial Response, Washington, DC, USA

USEPA (2000a) Electrokinetic and phytoremediation in situ treatment of metal-contaminated soil: state-of-the-practice. EPA/542. US Environmental Protection Agency, Office of Solid Waste and Emergency Response Technology Innovation Office, Washington, DC, USA

USEPA (2000b) Introduction to phytoremediation EPA/600/R-99/107. US Environmental Protection Agency, Office of Research and Development, Cincinnati, OH, USA

Uslu G, Tanyol M (2006) Equilibrium and thermodynamic parameters of single and binary mixture biosorption of lead (II) and copper (II) ions onto *Pseudomonas putida*: Effect of temperature. J Hazard Mater 135:87–93

Vaaramaa K, Solatie D, Aro L (2009) Distribution of ^{210}Pb and ^{210}Po concentrations in wild berries and mushrooms in boreal forest ecosystems. Sci Total Environ 408:84–91

Vandenhove H, Olyslaegers G, Sanzharova N, Shubina O, Reed E, Shang Z, Velasco H (2009) Proposal for new best estimates of the soil-to-plant transfer factor of U, Th, Ra, Pb and Po. J Environ Radioact 100:721–732

Vangronsveld J, Herzig R, Weyens N, Kristin JB, Ruttens AA, Andon TT, Erik V, Erika M, Daniel N, Mench VLM (2009) Phytoremediation of contaminated soils and groundwater: lessons from the field. Environ Sci Pollut Res 16:765–794

Vardanyan LG, Ingole BS (2006) Studies on heavy metal accumulation in aquatic macrophytes from Sevan (Armenia) and Carambolim (India) lake system. Environ Int 32:208–218

Veglio F, Beolchini F (1997) Removal of metals by biosorption: a review. Hydrometallurgy 44:301–316

Vishnoi SR, Srivastava PN (2008) Phytoremediation-green for environmental clean. In: Sengupta M, Dalwani R (eds) Procedding of Taal 2007:The 12th World Lake conference, Jaipur, India, pp 1016–1021

Vodnik D, Grčman H, Maček I, van Elteren JT, Kovačevič M (2008) The contribution of glomalin-related soil protein to Pb and Zn sequestration in polluted soil. Sci Total Environ 392:130–136

Walraven N, Bakker M, van Os BJH, Klaver GT, Middelburg JJ, Davies GR (2015) Factors controlling the oral bioaccessibility of anthropogenic Pb in polluted soils. Sci Total Environ 506-507:149–163

Wan X, Lei M, Chen T (2016) Cost–benefit calculation of phytoremediation technology for heavy-metal-contaminated soil. Sci Total Environ 563–564:796–802

Wang C, Fan X, Wang P, Hou J, Ao Y, Miao L (2016) Adsorption behavior of lead on aquatic sediments contaminated with cerium dioxide nanoparticles. Environ Pollut 219:416–424

Wang J, Shen Y, Xue S, Hartley W, Wu H, Shi L (2018) The physiological response of *Mirabilis jalapa* L. to lead stress and accumulation. Int Biodeter Biodegr 128:11–14

Wang JL (2002) Immobilization techniques for biocatalysts and water pollution contamination. Sci Press, Beijing

Wilde EW, Brigmon RL, Dunn DL, Heitkamp MA, Dagnan DC (2005) Phytoextraction of lead from firing range soil by Vetiver grass. Chemosphere 61:1451–1457

Wildt K, Eliasson R, Berlin M (1977) Effects of occupational exposure to lead on sperm and semen. In: Clarkson TW, Nordberg GF, Sager PR (eds) Reproductive and developmental toxicity of metals. Springer, New York, pp 279–300

Wuana RA, Okieimen FE (2011) Heavy metals in contaminated soils: a review of sources, chemistry, risks and best available strategies for remediation. Afr J Gen Agri 6:1–20

Xiong ZT (1997) Bioaccumulation and physiological effects of excess lead in a roadside pioneer species *Sonchus oleraceus* L. Environ Pollut 97:275–279

Yadav KK, Gupta N, Kumar V, Singh JK (2017) Bioremediation of heavy metals from contaminated sites using potential species: a review. Ind J Envir Prot 37:65–84

Yan G, Viraraghavan T (2001) Heavy metal removal in a biosorption column by immobilized *Mucor rouxii* biomass. Bioresour Technol 78:243–249

Yoon J, Cao X, Zhou Q, Ma LQ (2006) Accumulation of Pb, Cu, and Zn in native plants growing on a contaminated Florida site. Sci Total Environ 368:456–464

Yuan L, Zhi W, Liu Y, Karyala S, Vikesland PJ, Chen X, Zhang H (2015) Lead toxicity to the performance, viability, and community composition of activated sludge microorganisms. Environ Sci Technol 49:824–830

Zacchini M, Pietrini F, Bianconi D, Iori V, Congiu M, Mughini G (2011) Physiological and biochemical characterisation of *Eucalyptus* hybrid clones treated with cadmium in hydroponics:

perspectives for the phytoremediation of polluted waters. In:Book of Abstract, 5th European Bioremediation Conference. Technical University of Crete, Chania, Greece

Zaier H, Ghnaya T, Lakhdar A, Baioui R, Ghabriche R, Mnasri M, Sghair S, Lutts S, Abdelly C (2010) Comparative study of Pb-phytoextraction potential in *Sesuvium portulacastrum* and *Brassica juncea*: tolerance and accumulation. J Hazard Mater 183:609–615

Zhang W, Chen L, Zhang R, Lin K (2016) High throughput sequencing analysis of the joint effects of BDE209-Pb on soil bacterial community structure. J Hazard Mater 301:1–7

Index

© Springer Nature Switzerland AG 2020
D. K. Gupta et al. (eds.), *Lead in Plants and the Environment*, Radionuclides
and Heavy Metals in the Environment,
https://doi.org/10.1007/978-3-030-21638-2

—

Printed in the United States
By Bookmasters